T0343691

ZIMMER

ZIMMER

The MOVEMENT *That* DEFEATED A NUCLEAR POWER PLANT

ALYSSA S. McCLANAHAN

UNIVERSITY PRESS OF KENTUCKY

Scholarly publisher for the Commonwealth, serving Bellarmine University, Berea College, Centre College of Kentucky, Eastern Kentucky University, The Filson Historical Society, Georgetown College, Kentucky Historical Society, Kentucky State University, Morehead State University, Murray State University, Northern Kentucky University, Spalding University, Transylvania University, University of Kentucky, University of Louisville, University of Pikeville, and Western Kentucky University.

Editorial and Sales Offices: The University Press of Kentucky
663 South Limestone Street, Lexington, Kentucky 40508-4008
www.kentuckypress.com

Cataloging-in-Publication data is available from the Library of Congress.

ISBN 978-1-9859-0246-6 (hardcover: alk. paper)
ISBN 978-1-9859-0247-3 (epub)
ISBN 978-1-9859-0248-0 (pdf)

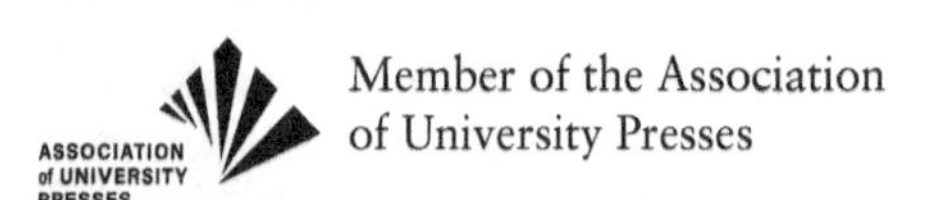

Member of the Association
of University Presses

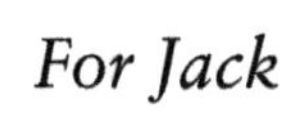

For Jack

Contents

Preface

This is a story of a deeply troubled nuclear power plant. It is a story of damaged and missing parts and of too much money expended. It is a story of a threat hanging in the air for years—the specter of radiation contaminating a community and the menace of billions of dollars of debt falling onto the shoulders of ordinary households.

It is a story of people: the people who built the power plant, the people who regulated it, the people who championed it. And the people who fought against it—to its death.

The story begins in 1969, when three utility companies in Ohio announced their intention to build a $395 million nuclear power plant in Moscow, Ohio, a sleepy town along the Ohio River twenty-eight miles southeast of Cincinnati. The story ends in 1984, when the resulting William H. Zimmer nuclear power station—with $1.7 billion invested into it—was canceled, 99 percent complete and still unlicensed by the federal government.

It was a dramatic fifteen years. By 1980, thousands of safety and construction errors had surfaced at the plant. Issues like junkyard steel being used for parts, welders with no welding training, the majority of systems installed incorrectly and with no paper trail. In 1981, the Nuclear Regulatory Commission slammed Cincinnati Gas and Electric, the company who owned the largest share of Zimmer, with a (then) record-high fine for a widespread breakdown in safety and quality assurance at the plant. And it got worse from there, implicating not only the utility companies in criminal mismanagement but also the Nuclear Regulatory Commission in poor regulation. As the project dragged on through the years—years marked by one of America's worst recessions and two energy crises—Zimmer's costs grew to be almost nine times larger than originally expected. And it was utility ratepayers who had to pay a portion of these wasted construction expenses in their monthly utility bills.

Through all of this, an evolving movement in the Cincinnati metropolitan area rejected the plant for both public safety and cost concerns. Each individual and group that emerged over Zimmer's lifetime framed public awareness in different and important ways, gradually enlarging the movement. Early on, only a few hundred people stood opposed; by the end of the 1970s, several hundred more joined them. By the early 1980s, as the plant's condition continued to worsen, the movement was twenty thousand strong. An odd collection of bedfellows—parents, local officials, teachers, scientists, utility ratepayers, homeowners, whistleblowing workers, and civil rights activists—these citizens gradually and temporarily united to oppose the Zimmer nuclear power plant. As much as they differed, they shared certain beliefs: that their communities deserved a healthy quality of life, that they had the right to voice their opinions, and that corporations and government agencies should be accountable to the public. They collectively questioned Zimmer's owners and its federal regulators, who—despite obvious signs of mismanagement at the plant—consistently ignored and sidelined any critique of the project, unilaterally forging ahead.

Ultimately, these forces caught up with Zimmer's champions. Local, dogged protest that persisted for fifteen years, combined with serious financial and regulatory pressures, compelled the utility companies to cancel Zimmer and convert it to coal. After thirty-five years as a (also controversial) coal plant, Zimmer finally closed its doors in 2022. Today, the plant sits empty and defunct, the shell of our story.

Preface

Zimmer came into my life on a muggy summer day in 2014. I was sitting with a friend, Phil Amadon, in the leafy Cincinnati neighborhood of Northside. Phil is a retired railroad welder and mechanic and a longtime labor activist; he was the secretary of his local union when I met him back in 2007. Through a mutual friend, Phil hired me to help him manage his work for the union. From that initial collaboration, we developed a lasting friendship. That hot summer day in 2014, when I was a grad student in history, I was telling Phil how I wanted to study activism in the past. He replied, "Did you know Cincinnati had a nuclear power plant that was stopped by local activists, myself included?" I was born and raised in Cincinnati but had never heard of a nuclear power plant there. A few weeks later, he handed me boxes of old papers—yellowed newspaper clippings, handwritten meeting minutes, phone trees, letters stamped out from typewriters. Everything was dated from the 1970s and the early 1980s, before I was even born.

From that, I dug in, slowly piecing together the saga of the Zimmer nuclear power plant. I scoured utility company archives, government hearing transcripts, press coverage, and records from the alphabet soup of organizations that formed against Zimmer. And while Phil was my starting point—my clue that Zimmer wasn't entirely welcomed in Cincinnati—I quickly learned that he and his coalition weren't lone voices, nor were they the first to question Zimmer.

The more I studied Cincinnati's failed nuclear power station, the more my drive to share its history hardened. Local protest alone did not single-handedly bring down Zimmer—but it was a crucial component. I couldn't shake this feeling that the people responsible for altering the course of Zimmer's history deserved to have their story told. This book tells that story—a story about ordinary people standing up to a powerful corporation and government agency, demanding a voice and accountability in the face of a looming disaster.

1

"More Economical Power"

CG&E's Nuclear Power Plant

On Friday, September 12, 1969, in Cincinnati, Ohio, a luncheon convened at Stouffer's Hotel in the downtown district. It was a warm, sunny day with highs in the seventies—"great football weather," as the *Cincinnati Post* commented. The hotel, a twenty-two-story concrete tower complete with a rooftop swimming pool, had just opened the previous year. It had been built to provide accommodations for the city's new convention center, which sat just across the street. The massive ninety-five-thousand-square-foot center, which took up an entire city block, was expected to attract hundreds of thousands of visitors each year and as such was surrounded by a sea of parking lots and garages. It even had a skywalk—an overhead pedestrian walkway connected to a garage—so that people would not have to walk on the streets after they parked their cars.[1]

Inside Stouffer's, a group of men sat in a conference room. They were city of Cincinnati representatives, local businessmen, and state and federal officials. In front of them stood a sixty-three-year-old man—his graying hair slicked back and firmly in place, his suit well cut—proudly announcing that Cincinnati had entered the nuclear era. His name was William H. Zimmer, and he was the retired president of Cincinnati Gas and Electric, CG&E for short.

He told his audience, "Following detailed engineering studies, the CCD group"—which the room understood meant CG&E, Columbus and Southern Ohio Electric Company, and Dayton Power and Light Company—"has decided it can further improve its electric system and produce more economical power with a nuclear fueled unit than by conventional coal-fired means. A nuclear power plant will help us put a brake on ever increasing costs in labor, equipment and fuel." CG&E, he explained, would build the plant along the Ohio River in the small town of Moscow, Ohio, twenty-eight miles southeast and upriver of downtown Cincinnati.[2]

CG&E had helped to build and promote Cincinnati's $10 million convention center. It was one of many efforts by the company to attract industry to the region and make it "an area on the move," as CG&E leaders liked to say. So, in this way, their decision to use the convention's hotel to announce the William H. Zimmer nuclear power plant (named after the CG&E executive himself) was fitting. The scene there, during the luncheon, also typified so much about CG&E leadership, which was characteristic of utility executives across the industry. Like William H., the other people in attendance were business and political elites who usually interacted only with other business and political elites. Among CG&E men there, the retired president and his associates had worked at Cincinnati Gas and Electric for decades, and in that time they operated with almost no public involvement in their decision-making: in their public announcement for a power plant designed to serve the greater Cincinnati area, there were no members of the public, aside from a few reporters, present.

And yet they saw themselves as longtime leaders and boosters of Cincinnati, for in their minds, energy development and metropolitan growth were tied together. Not only did they contribute to the city through the building of the convention center and other promotional activities, but they also supplied reliable and, most importantly, cheap power to the region, all the while providing a consistent return for shareholders. Over CG&E's long history—the company was over 130 years old when Zimmer was announced—it consistently grew its fleet of power plants, building ahead of demand and relying on technological innovation and economies of scale to make electricity quite affordable. CG&E leaders were proud of—and confident in—their ability to do this.

And so, when electric and gas sales and revenue skyrocketed after World War II during an unprecedented economic boom, when Americans were consuming electricity like never before, Zimmer and his colleagues felt assured in their decision to build a nuclear power plant. After all, nuclear energy had the ability to provide high-capacity, cost-efficient energy, making it—from the utility industry's perspective—a godsend for the postwar era. It also helped, significantly so, that the federal government's US Atomic Energy Commission (AEC), which licensed, regulated, and promoted nuclear power plants, was actively encouraging utilities to build them. And it helped that other federal agencies predicted electric demand would continue to rise precipitously in the postwar era. They recommended utilities diversify their sources of energy beyond just coal, oil, and natural gas, given these had growing supply and cost issues. Because of all these points, CG&E wasn't alone in ordering a nuclear power plant in the 1960s: so did a bandwagon of other power companies.

CG&E's decision, then, to build the Zimmer plant made sense at the time. Moreover, it was clear that CG&E leaders were trying (at least in their minds) to do their best. And yet it was also the case that they were foremost focused on costs and profits and maintained a deep-seated sureness in their company (and the way they had run it for decades). In these ways, CG&E executives were unwilling to be flexible, to change, as time passed. This, then, made them dismiss forces at work in American society—forces that would question if nuclear power plants were a prudent idea, what with the technology's risks for public health and environmental safety. These were silly qualms to CG&E directors because they trusted what "the experts" said. The men running the Atomic Energy Commission, and the men building the nuclear reactors at General Electric, and the men conducting research at universities and government labs—parallels to CG&E men in many ways—said that nuclear energy was, in the words of one nuclear engineering professor at the University of Cincinnati, "no threat whatsoever to the health and safety of the public—it's not a hazard, period." The professor then compared the possibility of an accident at Zimmer to the likelihood of the sun not rising tomorrow. It was with this company of men, and their collective confidence, that CG&E embarked on its nuclear power plant.[3]

Four decades after the city of Cincinnati was founded in 1788, CG&E was established. The city—situated along the Ohio River and nestled within steep hills—was then quickly becoming a—*the*, really—major metropolis and commercial port west of the Appalachian Mountains. By the 1830s, it had thirty-five thousand people and was growing. At the time, a minimal number of lard oil lamps supplied the only street lighting, so a group of public-minded boosters realized the city needed a dependable system of lighting. These men chartered the Cincinnati Gas-Light and Coke Company in 1837, and in 1841, local newspaper executive James Conover became its president and director, injecting the company with much-needed capital. In 1842, Cincinnati Gas-Light and Coke procured an exclusive gas contract with the city, built its first plant—which used distilled coal to manufacture gas for lighting purposes (and produced residual coke for heating)—and began to install gas mains for a system of city streetlights. In 1843, Cincinnati Gas-Light and Coke managers ceremoniously lit the city's first gas lamp at the W. H. Harrison Drug Store at Fourth and Main Streets. (In the 1920s, company officials nostalgically selected that same location for their new headquarters; they demolished the existing building and built a twenty-story office tower, which opened in 1930.) By the start of the Civil War, Cincinnati Gas-Light and Coke had expanded its services to 8,200 residential and commercial customers.[4]

With the advent of electricity in the 1880s, the company—then run by Civil War veteran Andrew Hickenlooper—built its first electric-generating station and received a franchise from the city, as did other local electric providers, making the market competitive. In 1889, Hickenlooper realized the potential to control both electricity and manufactured gas, so he bought up other electricity-generating companies. This was typical of American utilities, which, around the turn of the century, created monopolies in their service areas. By 1901, the Cincinnati Gas-Light and Coke Company was the largest energy provider in Cincinnati. It then changed its name to the Cincinnati Gas and Electric Company, known as CG&E.[5]

CG&E crucially lacked its own access to natural gas (which was cleaner, safer, cheaper, and more efficient than artificial gas), and so when another utility, Columbia Gas and Electric, completed a natural gas pipeline from West Virginia to Cincinnati in 1909, CG&E leaders made the difficult decision to become a subsidiary of Columbia Gas in 1911. Thereafter, CG&E invested in gas storage facilities around Cincinnati and expanded the city's gas system, connecting to other lines throughout the Midwest and the South. Gradually, by 1948, it phased out manufactured gas. In 1935, the Public Utilities Holding Company Act limited utility operations in an antitrust measure and forced Columbia Gas and Electric to divest many of its subsidiaries. CG&E, freed of Columbia, then purchased many of these, ironically consolidating power after Columbia's was diminished. Throughout the 1940s, 1950s, and early 1960s, CG&E continued to expand its gas and electric service area to the point that it became the main energy provider for southwest Ohio as well as portions of southeast Indiana and northern Kentucky. As another marker of its growth, it became publicly traded in 1946. By the end of the 1950s, it had around thirty-six thousand shareholders.[6]

From its founding, CG&E operated under growth-based development strategies, as did other US utility providers. Tasked by regulatory state commissions with providing continuous, reliable service to everyone in their service area, utilities had to raise enormous amounts of money to build power plants, which they did by selling preferred stock and fixed-rate short- and long-term bonds. (With common stock, a shareholder gets a percentage or share of profits, whereas preferred stock often has special voting rights and can be "preferred" in terms of payback, "preferred" ability to buy shares in the future, or other "preferred" terms. The difference between short-term and long-term bonds is that the former are paid back in the short term—within the next year or so—and have lower interest rates than those paid back in the faraway future, as in twenty or thirty years.) As one of the most capital-intensive industries, utilities built ahead of demand, relying on population growth data. And since it took a couple of years to build a power plant, construction—practically speaking—had to precede demand. Power companies also had a hand in creating demand, promoting an increased use of power. They lobbied homebuilders' associations to add greater electricity use into home designs. They advertised in newspapers, often with a cartoon character named Reddy Kilowatt who had a body made of red bolts of electricity. He reminded readers of how useful electricity was and that he never took a vacation. Power companies also targeted big industrial clients.[7]

And because power was cheap, people bought it. It was cheap because there were continual improvements in transmission and distribution and because manufacturers like General Electric kept developing technology that generated electricity more efficiently. It was also inexpensive because of the utility industry's business model—that as long as power companies built large, high-capacity power plants, the high upfront costs would be worth it in the long run. The companies produced more power that way, and the more they produced, the less it would cost for them to produce it. In other words, CG&E and others found that their unit costs decreased when they installed more power plants. Readily available fuel was another factor contributing to cheap power. For CG&E, it was coal coming from the mines of Kentucky and West Virginia. All of this meant that utilities could charge customers low rates, which spurred more demand. For shareholders, growing electricity sales meant increasing dividends. For power companies, it meant competitive stock prices, low interest rate charges on bonds, and continual investor interest. Everyone was satisfied.[8]

And because of utilities' ability to deliver affordable and efficient power, state commissions that regulated them accepted and supported power companies as "natural" monopolies. This created an industry culture with limited public involvement and powerful utility executives. Companies chose when and where to build new power plants in closed-door meetings with government and industry leaders, and they enjoyed a cozy relationship with local political leadership—in CG&E's case, with Republican city councilors, whose party dominated Cincinnati politics for the first half of the twentieth century.[9]

This was the utility industry's model prior to World War II, and it continued after the war, when unprecedented electric loads served as a major incentive for utilities to keep building. The war was a time of restrictions—ration cards, sacrifices, saving—so when it was over, many Americans had money to spend. The federal government encouraged this consumerism and subsidized economic growth, hoping to stave off another depression. It passed the GI Bill in 1944 to give veterans access to college tuition aid and home loans, hoping to grow the middle class. And it worked. After a brief postwar slump, the economy was booming like never before. In fact, Americans enjoyed the most sustained period of economic growth in world history from 1947 to 1973. With an abundance of well-paying, unionized industrial and manufacturing jobs, many American households in the 1950s and 1960s experienced an almost 50 percent increase in real median income, and at the end of the 1960s, many were wealthier than they were twenty years prior. Aided by federally backed mortgages, many households—albeit overwhelmingly white ones—moved to new single-family houses in suburban communities. Cheap electricity fueled this outward sprawl. Homes were wired, literally, for the latest available amenities, all of which required electricity: television sets, air-conditioning units, electric stoves, percolators, skillets, refrigerators, freezers, microwaves, washers and dryers, blenders, mixers, toasters, vacuums. The United States had become a high-energy society.[10]

In Cincinnati, these new suburban homes exploded along the two main highways that run north–south through the city and across the Ohio River, Interstate-75 and Interstate-71. Growth occurred on the far east, west, and north sides of Cincinnati; in northern Kentucky, new suburbs emerged across Boone, Kenton, and Campbell Counties. All in all, the area had around 1.2 million people by 1969. And as much as residential electricity use rose, so did commercial and industrial kilowattage, particularly in CG&E's suburban service areas, where industry was moving.[11]

As demand spiked in the postwar years, CG&E experienced unprecedented electric and gas sales and revenues. The utility's 1960 annual report shared, "In 1950, the average residential electric customer used 1,690 kilowatt hours of electricity; in 1960, he used more than double that amount." Over the next ten years, the average CG&E residential customer again doubled their electric usage. As such, in each year in the 1960s, the company reported record electric loads.[12]

Keeping ahead of the demand, CG&E rapidly expanded its arsenal of coal- and gas-fired power plants. From 1954 to 1964, it spent $338 million on construction and improvements and, anticipating future demand, planned to spend another $345 million from 1968 to 1973. In 1963, CG&E partnered with Columbus and Southern Ohio Electric and Dayton Power and Light for a power pool, where utilities combined resources to build large generating units. In 1964, that partnership resulted in a new, massive 434,000-kilowatt unit at CG&E's coal-powered Beckjord Station (named after another CG&E executive, Walter C. Beckjord, president from 1945 to 1957). Beckjord Station already had five other units in operation on its 1,500-acre site in New Richmond, a small town upriver from Cincinnati, close to where the Zimmer nuclear power plant would sit, but despite that large generating capacity, CG&E officials employed its industry's "economies of scale" logic:

build ahead of demand, then build big because the more you produce, the less it will cost you in the long run.[13]

By the late 1960s, CG&E was operating six gas- and coal-powered plants in the greater Cincinnati area, including a gas-powered station in the downtown district's old West End; three coal plants on the west side of town (close to the Indiana border); another in the village of New Richmond, close to Zimmer; and one farther east along the Ohio River, just opposite Maysville, Kentucky. These were joined by other utilities' electric power plants that sat along the river. Power stations stretched from Louisville, Kentucky, in the west to western Pennsylvania in the east. People, as the *Cincinnati Enquirer* and *Post* noted, started to refer to the Ohio River Valley as "power alley."[14]

These were good years for the men leading CG&E. Company portraits from the 1960s show a collection of older white men, all of them suited up—their hair white, graying, or thinning but always slicked back and firmly in place. Many, with their age, wore horn-rimmed glasses. Take William H. Zimmer, the namesake of the nuclear plant. Sixty-three years old at the time of Zimmer's announcement, he had worked for CG&E for forty-nine years, beginning in 1920 as a mail boy. He advanced to treasurer by 1943, a position he held until 1958. In 1945, he was elected to CG&E's board of directors and the following year became vice president. In 1957, he was made executive vice president and in 1962 president of CG&E. After his retirement, he served as a director with CG&E, which was his role upon Zimmer's announcement. Outside of work, he invested in and served on boards of several Cincinnati companies and institutions (the Bengals football and Reds baseball teams and the University of Cincinnati) and was a member of exclusive business clubs.[15]

His son, William H. Zimmer Jr., was the company treasurer, having worked there since 1952. The executive vice president, B. John Yeager, who in 1970 became the president, started at the company as a college co-op student. When CG&E publicized Zimmer in 1969, Yeager had been there for thirty-nine years. In

1951, Walter C. Beckjord, then-president of CG&E, wrote of his company, "There are actually now working with the Company a few employees who were born in Cincinnati before the Company entered the electric business. Some of them have been with the Company long enough to have been personally concerned with the Company's early dealings with Electricity." These lengthy careers created a company culture where higher-ups had deep loyalty to CG&E and great pride in their work (perhaps evident in how they named power plants after themselves). They boasted, accurately, that they worked for a utility skilled at building power plants (that is, *coal*- and *gas*-powered ones).[16]

They enjoyed their influence on Cincinnati beyond power generation. In an era of urban decline and growing suburbs (the city of Cincinnati lost 35 percent of its population in the second half of the twentieth century), CG&E executives saw themselves as helpfully stopping the hemorrhage. In the late 1940s, they launched a national advertising campaign designed to attract new industry to the city (which they assured people they could easily provide power for). In 1964, the company even formed its own real estate development company to work with city council and the local business community to promote revitalization of Cincinnati's downtown and riverfront areas. The work—which included parking garage construction; a riverfront stadium; office skyscrapers; the new convention center; and Stouffer's Inn, where Zimmer announced Zimmer—was urban renewal typical of 1950s and 1960s America. City planners and boosters—almost exclusively white men—redeveloped impoverished city centers toward new tourism and white-collar business in an effort to attract more tax dollars and more bodies back to old, declining city centers (even if just for a nine-to-five workday or a few-hours-long stint at the convention center). In their top-down planning, city leaders in the 1950s did not typically ask community members what they wanted; rather, leaders assumed they knew best.[17]

In 1963, to CG&E executives' delight, their company won a public relations award for its promotion of Cincinnati's downtown and riverfront redevelopment. This most likely meant little to customers, but to company officials, it was proof that the utility was a vital part of Cincinnati's growth and stature. "We have been more active than any other agency, and probably more than all agencies combined in the effort to attract *new industry* to the Greater Cincinnati Industrial area," CG&E president Walter Beckjord wrote in 1951. In annual stockholder reports, company officials relayed that CG&E's efforts to entice businesses back to the city's core "will be reflected in future sales and earnings."[18]

If these efforts didn't register with most customers, CG&E's ability to deliver cheap and dependable electricity did. And surveys in the years after World War II indicated that, as a result, CG&E's ratepayers were happy. Company executives relished that they were meeting consumers' high demand in the postwar years and that they controlled a system that produced benefits for everybody. Of course, utilities had been partly responsible for creating this demand in the first place, through masterful marketing. The president of American Gas and Electric, one of the nation's largest providers, told other utilities, "The most important elements that determine our loads are not those that happen, but those that we project—that we invent—in the broad sense of the term 'invention.' You have control over such loads: you invent them, and then you can make plans for the best manner of meeting them."[19]

Still, through the 1960s, CG&E officials expressed concern over their ability to provide reliable electricity into the next decade. While nearby coal was abundant—and CG&E continued to invest in it by planning multiple new coal-powered stations—coal costs were rising. Gas shortages were becoming an issue, too, and cheap oil did not solve the problem. Utilities began using larger quantities of oil in the 1960s to generate electricity in times of peak demand (although CG&E relied on coal for 95 percent of its power production). In the midst of these supply issues, Americans' insatiable electricity use was stretching electric grids to peak capacity, resulting in power outages. One blackout—on Tuesday, November 9, 1965—plunged twenty-five million people in New York City into darkness for over twelve hours. In 1967, to prevent such an event, CG&E and twenty-two other utilities in eight nearby states signed an agreement to share power plant operations and transmission line data to prevent "a mammoth blackout," as William H. Zimmer, then-acting president of CG&E, stated at the time. It wasn't just utilities that expressed concern over growing electricity needs and related power shortages. Federal agencies, like the Federal Power Commission and the Bureau of Reclamation, did too (beginning in the 1940s), and they encouraged power companies to identify diverse sources of energy.[20]

So, as CG&E executive William Dickhoner captured in 1971 when he said, "Our sole remaining energy source as we know it is uranium," CG&E turned to nuclear power. But while CG&E did so in part to avoid its issues with coal and natural gas, the main justification for Zimmer was its ability to supply high-capacity, cost-efficient energy—or as William H. Zimmer called it in 1969, nuclear energy's "economical power." (Interestingly, this was also why utilities built massive coal power plants like CG&E's Beckjord station. Indeed, coal and nuclear weren't mutually exclusive; utilities including CG&E pursued both simultaneously.) Uniquely, though, nuclear power plants could continuously generate enormous power from a small amount of uranium fuel. In 1969, when CG&E executives announced Zimmer, they told shareholders that it would eventually provide power to one-fourth or even one-third of all its customers.[21]

Still, it took CG&E a minute—actually, most of the 1960s—to reach its decision to build a nuclear power plant, a timeline it shared with many other power companies. While the cost of nuclear fuel was cheaper to mine, process, store, and transport than coal, the capital costs to build a nuclear station were significantly higher than for coal-powered or natural gas facilities. Operational and maintenance expenses were also comparatively high, not to mention there were decommission and waste disposal fees. CG&E and the other utilities needed some help, then, some incentivization to get through the initial—and very expensive—investment stage with nuclear power. Luckily for them, the Atomic Energy Commission—the AEC—was there to help.

The 1946 Atomic Energy Act established the AEC as a federal agency to exclusively manage the production, ownership, and use of fissionable material for the nation's defense. In 1954, another Atomic Energy Act permitted utility companies to produce commercial nuclear power. But even before that, utility companies got their feet wet in the technology by providing power for national defense facilities involved in nuclear technology. In Ohio, power companies pooled resources in 1952 to build two new coal plants in Portsmouth (upriver from Cincinnati), which then fueled the construction of a gaseous diffusion plant producing enriched uranium. Additionally, as scientists from universities and government laboratories were experimenting with different kinds of nuclear reactors for civilian electricity, CG&E and other utilities pooled their resources and sponsored some of this research and development. This familiarized many private utilities with the basics of atomic energy. Still, the high capital costs of nuclear power stations intimidated companies like CG&E from ordering reactors themselves for much of the 1960s. By 1962, only six private nuclear power plants were operational in the United States.[22]

In stepped the AEC, which treated the fledgling nuclear power industry with sympathy. It was the early Cold War, and as a part of that, President Dwight Eisenhower broadened the definition of national security to include US technical superiority and economic strength over the Soviet Union. Commercial nuclear power was one project to meet this goal. To this end, Congress tasked the AEC with both the promotion *and* the regulation of this new industry, in the hopes of getting nuclear energy off the ground. In practice, officials at the AEC were much more consumed with the first task. To be fair, they were under a lot of pressure from their congressional oversight committee, the Joint Committee on Atomic Energy, to encourage the growth of nuclear power. And beyond that, many in the Atomic Energy Commission and its associated bodies were individuals with long careers in national defense. Many had served in the military, and others had worked at government-run or -backed laboratories involved in nuclear research. As such, they were deeply enmeshed in and faithful to their line of work, and since they were working on research related to Cold War national security, they were used to insular—secretive, really—decision-making. All of this made their agency's culture in many ways similar to CG&E's.[23]

With AEC promotion, the tide turned for commercial nuclear power. In 1957, Congress passed the Price-Anderson Act after lobbying from members of the Joint Committee on Atomic Energy. This act—which limited the liability of companies operating nuclear plants in the event of an accident—was vital for the young nuclear power industry. Prior to it, the insurance industry couldn't provide adequate accident coverage for private utilities operating nuclear power plants. By limiting power companies' liability, the act was in effect a permission slip for CG&E and others to give nuclear power "a try" with fewer financial risks. There were other helpful developments. Until 1964, the Atomic Energy Commission retained ownership of nuclear fuel and leased it to operators of nuclear power reactors at subsidized rates. (The AEC lobbied to end the subsidy only when it felt that nuclear power could compete against fossil fuels.) General Electric and Westinghouse, the major builders of nuclear reactors, also started to construct turnkey plants—nuclear stations built for a set price—which subsequently placed financial liability on these builders. Power-pooling arrangements, which CG&E had done for its coal-powered Beckjord unit in 1964, also helped immensely. Since nuclear power plants were so expensive to build, utilities teamed up to construct them, each contributing funds. Furthermore, power companies became comfortable—and confident—with the associated learning curve of nuclear power. Many believed that as they became more experienced, labor costs would be later held down, saving money.[24]

By the mid-1960s, construction costs for nuclear power plants were more manageable or at least seemed more manageable.

Between 1966 and 1968, utility companies purchased sixty-eight nuclear reactor units. By 1972, twenty-nine nuclear power plants were in operation or licensed, with fifty-five under construction and seventy-six planned. Many were located on the East and West Coasts, but there was also a concentration of planned, ordered, or under-construction nuclear power plants in the Midwest around the Great Lakes: in 1967, the Northern Indiana Public Service Company ordered the Bailly nuclear power plant along the shore of Lake Erie in northern Indiana, and three years later, the Toledo Edison Company and Cleveland Electric Illuminating Company broke ground on their Davis-Besse nuclear power station near Toledo, Ohio. Others in northern Ohio, Indiana, and Illinois and in southern Michigan would follow in the early and mid-1970s. Similarly, the Ohio River was another Midwest location ideal for nuclear power stations. Aside from CG&E's Zimmer, Public Service Indiana and its partners decided to build a nuclear power station in Madison, Indiana, about seventy miles southwest of Cincinnati, along the Ohio River. By the early 1970s, the federal government projected that by the decade's end, almost one-fourth of US electric capacity would come from nuclear power, rising to around 50 percent in 2000.[25]

All of these developments pushed CG&E toward Zimmer. In the mid-1960s, its executives, along with those at Columbus and Southern Ohio Electric and Dayton Power and Light, consulted with an engineering firm for their next power-pooling project, and in 1969, they decided nuclear power made the most "economical" sense. They ordered two reactor units for Zimmer, anticipated construction on the plant site to start in 1971, and expected both reactors to be operable by 1976. Beyond that, CG&E leaders noted the possibility of three additional nuclear power plants for southwest Ohio. Since Zimmer sat in CG&E's service area, and not in Columbus or Dayton's, CG&E owned the largest share of the plant. Columbus and Southern Ohio and Dayton Power owned 28.5 and 31.5 percent of the plant, respectively, and were entitled to that percentage of energy, which would travel through power lines from Moscow, Ohio, north and northwest.[26]

Producing nuclear energy is a complicated technology to say the least, but at the most basic level, it boils down to using water and fission to produce power. First, there's the reactor. CG&E and its partner utilities chose a General Electric boiling water reactor, designed to hold 115 tons of uranium fuel. It would sit in Zimmer's main structure, a large building called a reactor-turbine complex, which was made of reinforced concrete and structural steel, giving it a practical, no-frills appearance from the outside. The complex connected the reactor to a turbine, generating electricity through nuclear fission. Purified water—sourced from the Ohio River—would enter the bottom of the reactor core, flowing upward around cylindrical fuel rods containing uranium dioxide fuel pellets. To produce electricity, neutrons would collide with the uranium inside the core, causing atoms to split. The generated heat and energy would boil the water surrounding the core, creating steam. That steam would then travel through a series of pipes to a turbine generator, powering it. The spent steam would run to a large surface heat exchanger, where it would cool, condense, and return to the reactor vessel. The cooling tower—a giant hourglass as tall as a forty-story building—helped with removing leftover heat. It received 450,000 gallons of condensed water per minute, which it circulated over concrete-asbestos sheets at its base. Excess heat would rise by means of natural convection, exiting the top of the tower (a process that was further aided by a current of air that traveled vertically through the tower). Near the cooling tower, there would be an office, a storage facility, a pumphouse, a sewage treatment plant, and barge docking along the Ohio River. Five hundred feet inland from the reactor building, there would be a switchyard, where transmission lines would carry power to a new substation.[27]

CG&E directors understood how nuclear power functionally worked. Like other utility heads venturing into nuclear power around 1970, they reminded people of their familiarization with and contacts in the emerging industry, using that to justify that they knew how to build and operate Zimmer. As part of their construction application to various federal and state bodies, company representatives wrote that "the architect-engineer and reactor supplier [for Zimmer] have had extensive experience in nuclear power plant design." They noted that several of their engineers had worked in nuclear weapons production laboratories and on reactors, and dozens of others had taken courses on nuclear power. CG&E's engineering consultant, Sargent and Lundy, had designed nuclear reactors since the mid-1950s. CG&E also chose a reactor from General Electric, which had been making them since the mid-1950s.[28]

Hiring these kinds of experienced firms allowed CG&E officials to show their customers, shareholders and investors, government reviewers, and partner utilities in Columbus and Dayton that they knew what they were doing. To have not done so would have meant difficulties in financing and permitting. But it was also that CG&E leaders—like many other utility heads—*really* believed in their ability to execute a nuclear power station, despite having no experience with the financing or the permitting or the construction, the construction management, the government regulation, and the operational maintenance entailed in running a nuclear power station. To this point, CG&E selected the Henry J. Kaiser Company to be Zimmer's general construction manager because, in the words of CG&E executives, Kaiser "has provided construction and consulting services to the Atomic Energy Commission continuously since 1950." In reality, Kaiser had overseen the construction of *two* experimental reactors for the federal government, including one at the Hanford, Washington, nuclear production complex—and that was it. So while Kaiser could boast of other substantial construction projects, including the Hoover Dam, it had no experience managing the construction of a nuclear power plant. And yet CG&E leaders optimistically anticipated that Kaiser would still get the job done.[29]

State Route 52 hugs the Ohio River as they both trace the border between Ohio and Kentucky. Driving along it, going eastward from the city of Cincinnati, you notice rural farmland, dense pockets of large overhanging trees, the occasional bend in the road, the occasional incline or decline, power lines racing above and alongside your car. The river comes in and out of view. You see signs for other old state routes, arrows indicating turnoffs for small riverside towns. Zimmer's hometown, Moscow, is one of these. Five blocks wide, from Water Street to Fifth, it is a tiny hamlet. When CG&E announced Zimmer in 1969, Moscow had just shy of five hundred residents. There were two grocery stores, one elementary school, a town bar, a machine shop, a garage, a carryout restaurant, and historic homes dating back to the early 1800s. There was no stoplight because there was no real need for one.[30]

For Zimmer, CG&E selected a six-hundred-acre site just north of Moscow's "downtown," on land it had purchased in the 1950s, then intending it for another coal plant. But since the location's features were ideal for a nuclear power plant, the utility decided to use the land for just that. Like Moscow, the land Zimmer sat on was flat—great for construction. Beyond that, it was right next to the Ohio River, ideal since nuclear power plants require significant amounts of water and (some of) Zimmer's uranium fuel was going to arrive via river barges. The area did not typically experience severe weather like earthquakes. The worst case was a tornado or a flood. The latter was a real possibility given that a major flood in 1937 had devastated Moscow—and pretty much every other town along the Ohio River. To compensate for this, CG&E built three feet above the historical flood record. CG&E's selected site was also close to other substations, transmission lines, and transportation, including the Chesapeake and Ohio Railroad paralleling the river on the Kentucky side, making access to Zimmer easy.[31]

Another reason CG&E picked Moscow was because it and the surrounding area was a rural "low-in-population zone," called an LPZ. The biggest town in the area—New Richmond—was six miles from Zimmer and had only 2,650 residents. Zooming farther out, Zimmer sat in Clermont County, the county to the immediate east of Hamilton County, where the city of Cincinnati was.

While Clermont County was growing with new suburbs, most of it, particularly near Zimmer, was (and is) rural farmland.[32]

In the 1960s, when utilities were building nuclear power plants across the United States, many were sited in LPZs, reducing the scale of impact if something went wrong. This was not just utilities being overly cautious: the Atomic Energy Commission encouraged, and in many cases required, LPZs for nuclear power plants. After the commission permitted the commercialization of nuclear energy in 1954, its officials developed site guidelines, favoring non-dense areas. But because utilities pushed back, concerned about the cost of transmitting electricity over long distances from rural areas to population centers, the AEC said it would consider conservative reactor design as a mitigating factor against an urban location. So, according to the commission's criteria, if a power company could show that a nuclear power plant located near a population center was designed with multiple, redundant safety features, the commission *might* permit its location. That said, commission officials basically disregarded their own provision—as was made apparent in the early 1960s, when the utility company Con Edison unsuccessfully tried to site a nuclear power plant in Queens, New York City. Instead, the AEC—from then on—forced utilities to put their nuclear power stations in rural and suburban areas.[33]

CG&E, in its permit to build Zimmer, acknowledged these rules, explaining to government authorities its decision to put Zimmer in rural Clermont County. By doing so, CG&E officials conceded, just like Atomic Energy Commission officials did, that nuclear power was not risk-free—that there *could* be an accident, releasing a large amount of radiation—and that in everyday usage, a small amount of low-level radiation would be emitted from Zimmer. To receive a construction permit and an operational license from the AEC, CG&E had to study both of these realities, estimating radiation levels for high- and low-level releases. But CG&E leaders did not believe these risks amounted to actual threats and, in the words of one executive, the "benefits gained by our society as a whole far exceed the risks." He said Zimmer would emit casual radiation "only in amounts that are well below what can be considered harmful"—levels that were "practically zero." And he likened the chance of a nuclear malfunction at Zimmer to that of an airplane crash—possible, yes, but statistically unlikely, to the point that the Cincinnati area *could* and *should* ignore the possibility. So CG&E's choice of Moscow because it was an LPZ was lip service to an idea of risk that company leaders didn't buy into.[34]

Convinced of Zimmer's safety, CG&E executives liked to remind Moscow and other nearby towns that, in building the nuclear power plant, it was *not* adding another coal station. At Zimmer's announcement in 1969, B. J. Yeager, then-vice president at CG&E, said that "compared to what fossil fuel plant puts out," Zimmer would not pollute the air. The Ohio River Valley had horrible air quality, thanks in large part to CG&E and other utilities' use of nearby high-sulfur coal and how it hung over the valley. To this point, some residents—instead of referring to the region as a power alley—derisively called it pollution alley, as *Cincinnati Post* reporter Douglas Starr captured. Coal pollution was so bad that housewives complained of having to daily scrub soot from their walls and windows. Even lichen disappeared from the region for a time (and only reappeared in the twenty-first century).[35]

So Yeager's point had salience with the community around Zimmer. One Moscow resident told the *Cincinnati Enquirer*, "I'd rather see an atomic plant than one [a coal-powered plant] like at New Richmond where all the smoke comes out." Americans had been paying more attention to air quality since the turn of the twentieth century, when industrialization filled the skies with dark soot. But after World War II, when industry was booming, the issue became more pressing, especially after disasters like that of Donora, Pennsylvania, in 1948. There, smog became so thick and dense it was literally deadly, killing twenty people.[36]

Playing up Zimmer's "clean" energy, CG&E's 1969 renderings of the power plant showed Zimmer surrounded by green grass, tall trees, and clear blue skies. Its cooling tower, in contrast to the smokestacks throughout the region, was *not* generating a plume of smog. And yet—considering how CG&E fought new air pollution legislation all through the 1970s and how it continued to invest in coal-powered stations—the impression that its leaders were concerned about air quality was quite disingenuous.[37]

In 1970, the year of America's first Earth Day, Republican president Richard Nixon created the Environmental Protection Agency, the EPA, and signed into law the National Environmental Policy Act, usually called NEPA, which required federal agencies to assess the environmental effects of their proposed actions prior to making decisions. At the same time, the Clean Air Act Amendments of 1970 empowered the EPA to set national air quality standards determined by the best available science, and states had to come up with implementation plans to be approved by the EPA, which Ohio's Air Pollution Control Board did. New point sources of pollution—those that are singular and identifiable, like coal smoke pouring out of a smokestack—suddenly had to have a federal permit that included approval of the technology used to control emissions. Altogether, the 1970s saw unprecedented federal and state reviews of stationary sources of air pollution.

CG&E executives met this with fierce opposition. They told shareholders and the press that the mandates were "excessive," "extremely difficult, if not impossible to achieve," and "rigidly enforced." To be fair, new emission standards *did* prove very costly for CG&E and other coal-dependent utilities. In 1971, one year after NEPA's passage, CG&E leadership wrote its stockholders, "Our ability to meet air standards on sulphur dioxide is going to cause us problems"—meaning a greater financial burden to implement new smokestack-screening technology to rid the coal smoke of certain particulates. Coal from Kentucky and West Virginia had a 3 percent sulfur content, but new standards called for coal to have 1 percent or less. Mines west of the Mississippi River had such coal, but transporting it to Cincinnati would "double the price of coal," CG&E officials wrote shareholders. Conversion to natural gas was not a solution since the gas supply was tightening. In 1972, CG&E's natural gas supplier placed restrictions on the availability of increased supplies of natural gas for the future.[38]

CG&E's solution to this problem was to file a suit against the Air Pollution Control Board for its new standards, which it did with other Ohio utilities. Ultimately, though, Ohio EPA forced utilities to comply with particulate emission limitations by 1977. Nonetheless, it was revealing that CG&E acted just like the automobile, lead, rubber, and chemical industries at this time that similarly argued that federal and state regulators were asking for extraordinary technological innovation on the part of companies to meet new environmental regulations. These industries spent substantial resources on proving there was no *definitive* link between their product's use or manufacture and people's safety. They also asserted that, under certain thresholds, concentrations of chemicals and particulate matter were trivial for human bodies and, since the EPA was a new bureaucracy, its public health studies justifying environmental regulations should be considered inconclusive. (Cigarette companies said the same to defend their products.)[39]

All of this revealed that CG&E was not actually *that* interested in improving the region's air quality, just like it was not *that* worried about Zimmer's public health ramifications. Conversely, its leaders *were* worried about costs. And so, as air pollution legislation began to expand in and after 1970, the company's commitment to Zimmer solidified. CG&E officials did not think nuclear energy would come under the same extensive and costly environmental fixes that coal plants were enduring. (And this proved true—although only for the briefest glimpse of time. After a 1971 court decision, NEPA's environmental review process suddenly applied to nuclear power stations.)[40]

In its promotion of Zimmer, CG&E also emphasized the tax revenue and thousands of jobs that the plant would generate for the surrounding economically depressed area. Moscow had always been a tiny town, but it had certainly fallen in prominence over the last hundred years. Founded in 1816, it was a vital stop on the

Underground Railroad, like many other Ohio riverside towns, and before the 1920s was a major producer of brandy. Prohibition killed that, and the 1937 flood caused significant property damage. Even after most American roadways were paved, many of Moscow's were gravel. Some people, as late as the 1960s, still used outhouses.[41]

By the mid-twentieth century, as cities and towns through the Midwest and the Northeast lost industrial and manufacturing jobs to more favorable tax locations, Moscow joined them in suffering from a stagnant economy with few nonagricultural job prospects. While the area around Moscow was never a major industrial center, it nonetheless felt these trends. In the early 1970s, unemployment was around 14.5 percent. Aside from a surgical supplies company and two quarry operations, there weren't many industrial or manufacturing jobs around. High-paying positions at New Richmond's Beckjord coal plant provided livelihoods for some Moscow families, but many had to commute into the city of Cincinnati for factory work. Some men drove trucks for a living or were construction workers. The remaining families farmed. Corn, tobacco, and alfalfa were common to the area as were hay, soybeans, wheat, oats, and vegetables. Within fifteen miles of the Zimmer site, there were three hundred dairy farms.[42]

So, with the news of Zimmer, many residents, businesses, and elected officials in and near Moscow couldn't help but be elated with the project. After all, Zimmer was estimated to double Clermont County's tax duplicate (real estate assessments on which taxes are based) by the time it opened in 1976. "With Zimmer's announcement," one resident recalled, "people near there thought the streets would be paved with gold." In this elation, Moscow was like other small towns across America during the Cold War that, largely, embraced having a nuclear power plant or nuclear weapons production site near them for the middle-class affluence these facilities brought. Two years after CG&E's announcement, Bob Lynn of the *Cincinnati Enquirer* interviewed a handful of Moscow residents. The hope of Zimmer's economic stimulus was still in the air. James O. Foster, the soon-to-be mayor of Moscow, told Lynn, "Moscow could become one of the nicest villages along the river. We could have parks, new schools, sewers."[43]

Several of those interviewed said they weren't that concerned about the possible health risks associated with being near nuclear technology and assumed the federal agencies in charge of nuclear power had public safety in mind. After all, the Atomic Energy Commission and its associated bodies included and involved highly trained nuclear engineering experts—which, to many Americans, meant they, and thus the AEC, were trustworthy. Moscow's elementary school principal said to Lynn, "I'm not afraid of it. I'm not moving. People are for it. We're tickled to death to have it. It'll put Moscow back on the map." One woman who had lived in the community for forty years was also hopeful: "I wonder if it'll help Moscow. We sure need some help." Another Moscow resident man said, "I think the plant's a good thing. They"—meaning the government—"must know what they're doing or they wouldn't let them do it." His neighbor echoed him: "I don't think the higher law would allow them to put it here if it wasn't safe. Do you think they'd just put anything up that'd kill you? I think it'll be a good thing for the town and my kids." Another resident said the same: "I've heard they"—again, the government—"check every plant and what comes out."[44]

The following year, in 1972, Foster, then Moscow's mayor, said publicly, "I live and run a small business in Moscow Village within one-half mile of the proposed William H. Zimmer Power Plant site. I have no fear concerning the safety and well-being of my family or the people in this community. In my opinion, with the technical knowledge available today, I feel sure that the Cincinnati Gas & Electric Company is doing everything possible to provide a safe power source." A village councilman from New Richmond, close to Zimmer, likewise stated that his constituents wanted the plant: "New Richmond will be the closest community to the Zimmer station. I feel the majority of people here and in Moscow are in favor of the plant."[45]

Aside from jobs and tax benefits, local and county officials agreed with CG&E that it was imperative for southwest Ohio to have additional power stations to manage its growing population and commercial and industrial development. The Clermont County board of commissioners insisted that delays in power station construction and licensure "will increase the cost of electrical power to the users" and further hurt the economy of Clermont County. The mayor of another small town near Zimmer echoed this, saying in 1972, "I think we should . . . follow the guidelines of the Atomic Energy Commission, and correct our condition which is potentially a lack of power. We don't want blackouts. We don't want brownouts." The board chairman of Armco Steel, a major steelmaking company that had a large production facility north of Cincinnati, likewise insisted that Zimmer—the power it would supply—was necessary to keep his company in the area.[46]

Still, regardless of Zimmer's economic and energy benefits, there were Moscow-area residents who were unsure of the project. Bob Lynn at the *Enquirer* summed it up in 1971: "It seems the 486 souls in Moscow are either losing sleep worrying about having an atomic plant in their backyard or are staying awake just thinking about all the wonderous things such a plant would mean to the village." More to the point, "some Muscovites think it is a good thing, some don't." One woman who owned a gas station along State Route 52 with her husband told Lynn, "I'm kinda dubious about the atomic plant. You hear so many things you just don't know what to believe. Everyone who drives into the station asks us if we are worried about the plant. But I guess we just have to trust the government." Another resident, concerned that Zimmer was too close to their house, said, "I'm going to plant a row of pine trees in my yard and block it out of my view." One mother told Lynn, "I thought it was all right until the other day when a doctor on the Nick Clooney TV show said that if it blew up it'd kill us first, so we wouldn't have to suffer."[47]

These statements of unease from 1971—about Zimmer's public health impacts, about how much people could trust the government—foreshadowed the opposition to the project that would soon emerge. In fact, only the following year, during the first government hearings held on Zimmer, people began to object to the plant. CG&E leaders weren't obtuse to this, but they didn't take the threat of criticism and protest seriously. Instead, they had plans drawn up for a visitors' center on the Zimmer site, indicative that they thought people would want to travel to and marvel at the plant.

2

"Every Single Dose of Radiation Carries a Finite Risk"

Protest Begins

On Friday morning, May 12, 1972—a cloudy spring day—a biology and chemistry professor named David Fankhauser stood in front of a trio of seated men at the Clermont County courthouse. The two-story red-brick building sat on the main street of Batavia, Ohio, a small town with fewer than two thousand people, about twenty miles north of the Zimmer nuclear power station. Fankhauser was there for the first public hearing on the power plant—and it was not going well for him.

The three men in front of Fankhauser were members of a federal Atomic Safety and Licensing Board, and they were evaluating CG&E's application for a construction permit. Fankhauser, who lived near the proposed Zimmer plant, was telling the board that he was very concerned about it. Specifically, he was detailing what a serious accident could look like. John B. Farmakides, the board chairman, had just interrupted him, asking, "Do you affirm that the statements that you are making are, in fact, true?" Fankhauser, a lanky man who usually wore glasses, replied, "I am saying them."

Farmakides pressed him: "Do you understand what it is to say them to a governmental body? Are you aware of 18 USC 1001, what this says? Let me paraphrase it for you. Because what you are doing is providing information. I think you should be aware of the fact that you are providing it to a governmental body . . . if you make any false statements or any misinterpretations to this body, the Board, you are liable to fine and imprisonment. I would like to know if you are aware of that."

Fankhauser quietly responded, "Well, I am certainly aware of it by now."[1]

Now, David Fankhauser knew how to protest. Raised in a Quaker household by a mother active in civil rights, pacifism, and antinuclear protest, he attended college in the early 1960s at Central State University, a historically Black school near Dayton, Ohio. There, the Congress of Racial Equality, a civil rights organization, recruited him to be a Freedom Rider. As part of that, Fankhauser and others traveled to Alabama and Mississippi on buses, fighting violently enforced (and unconstitutional) racial segregation in Southern interstate buses and bus terminals. In his civil disobedience, he worked under legends like Ralph Abernathy and even Martin Luther King Jr. Following one demonstration in Jackson, Mississippi, in 1961, where Fankhauser and other white and Black students sat down in a white-only bus waiting room, they were arrested and refused bail. (Fankhauser and another student from Central State were the first two white Freedom Riders to do this in the 1960s.) The group spent forty-two days in the Parchman State Penitentiary in deplorable conditions before civil rights groups secured their release.[2]

Yet, even with this background, after the man had faced considerable violence in the Deep South, Fankhauser felt intimidated by the Atomic Safety and Licensing Board and by CG&E officials. As the first hearing on Zimmer established, most of the battle over the plant took place in conference rooms and courtrooms, where public meetings were convened. And as Fankhauser and Farmakides's interaction showed, public hearings were not that inviting to the actual public. Scientific and legal experts with ties to government-backed nuclear research staffed Atomic Safety and Licensing Boards, and most members were known, like Atomic Energy Commission officials, to be hostile to public opinion. (Even the physical arrangements of the hearings favored CG&E and government representatives: officials sat behind desks at the front of the room, while members of the public had to stand or sit at the back of the room and were only called forth when given a chance to speak. And then they approached officials with their backs to the audience.)

Still, Fankhauser persevered, and he was not alone. In fact, he and around one hundred others came to Zimmer's first hearing in May 1972. The licensing board commented it was "quite a large number of people to handle," and they had to move rooms in the courthouse to accommodate everyone. The next meeting, around the same number showed up.[3]

These hearings were a required first step for CG&E to build and operate Zimmer. To operate a nuclear power plant, a power company had to first apply for a construction permit from the Atomic Energy Commission (which CG&E did in 1970) and wait for approval before it could proceed with the majority of construction. Then, when building the plant, the utility applied for an operational license to provide power. While it was the commission that awarded the permits, prior approval was required from other federal and state agencies, too, including a federal Advisory Committee on Reactor Safeguards and an Atomic Safety and Licensing Board, both of which were staffed with technical experts who provided additional, and independent, feedback to AEC officials on reactor facility license applications. So it was at this May 1972 meeting that a licensing board was first reviewing CG&E's proposal to build Zimmer. Up for discussion was Zimmer's environmental impact during construction as well as the plant's safety and design features—public information that had been included in CG&E's application for a construction permit. Members of the local community were permitted to speak—that is, in no more than five-minute intervals. This was because they weren't considered "full" participants of the proceedings.

Fankhauser's protest at the 1972 hearings—and his treatment by the Atomic Safety and Licensing Board—marks the start of the protest against Zimmer, not just because he was one of the earliest to say something but because he had so much in common with other early protesters, and together they formed an initial wave of opposition. Most of them owned a home near Moscow, Ohio, with land or a farm. Many were parents. Fankhauser checked all these boxes, although unlike most of his neighbors, he also had a PhD. After college, he attended John Hopkins University for a doctorate in genetics. There, his research focused on radiation-induced mutations in bacteria, giving him high-level knowledge of radioactivity's implications. There were others at the hearings, though—heads of local environmental groups—who similarly wielded PhDs and activist experience.

Despite the difference in academic credentials, the same concerns united people in the audience: What would Zimmer do to the land, to the water? Some people asked: Was another power plant needed? Above all, community members worried about human bodies, especially their children's. Fankhauser and his wife, Jill, had just moved to Moscow's neighboring town, New Richmond, the year before, not too far from where he taught chemistry and biology at the University of Cincinnati-Clermont. They rented a stretch of land because they wanted to raise their family in a sustainable, off-the-land way. In their minds, Zimmer jeopardized their ability to do that. Like any nuclear power plant, Zimmer would emit low-level radiation in everyday operations and in the event of a major accident could release a large amount of radiation. Both scenarios troubled the Fankhausers and several of their neighbors, especially since certain safety features in nuclear power stations were undergoing fresh government review.

In their unease or outright rejection of Zimmer, the families who lived near the plant were thinking of themselves since they lived in its shadow, but they also told the Atomic Safety and Licensing

Board they didn't want the nuclear power plant moved elsewhere in the region. They were concerned about nuclear power for everybody, everywhere. This reflected how more Americans were taking seriously public health and safety by the 1970s as environmental awareness spread. They were doing their own research into topics like nuclear power, along the way critiquing government-backed scientific views from authorities like the Atomic Energy Commission. People's gumption to do this—to get a copy of CG&E's construction permit, to try to make sense of it, and to then challenge government and CG&E officials about information in it—also showed how Americans expected more accountability from and accessibility to their leaders by the 1970s. After decades of political and business elites running the show, ordinary citizens took momentum from civil rights and community-organizing activism and began demanding a say in local decision-making. To this point, several people, including Fankhauser, petitioned the AEC that they had a specific interest in the case of Zimmer (like being a homeowner near it), and as such they should be an intervenor in future hearings, giving them the right to fully participate in the proceedings and to affect change that way. It would take them four years to gain that status, and in the process, licensing board chairman John Farmakides would harangue David Fankhauser more than once.

On Friday, May 12, 1972, during that first hearing on Zimmer, David and Jill Fankhauser began the public portion of the meeting by telling the Atomic Safety and Licensing Board that Zimmer wasn't even necessary in the first place. Jill said, "We comprise 1/16 of the world's population but are using 1/3 of the world's power. I would like to also mention the amount of power we use now seems in excess to me from the point of view of Clermont County and looking around and seeing the fantastic quantity of smoke pouring into the air from the Beckjord [coal] Plant." She suggested that they open the courthouse's window curtains and turn off the electric lights to save energy. Natural light worked just fine, and the courthouse had such large windows.[4]

In the early 1970s, Americans felt a growing sense of environmental crisis—and a growing sense of responsibility for it since many of the issues stemmed from technological change and economic growth. There was so much smog in Los Angeles from sprawl and cars that people wore gas masks. The Cuyahoga River in Cleveland was so polluted from nearby industry and waste that it caught fire. And as much as American abundance had created environmental problems, urban and rural poverty did too. In inner cities, lead paint poisoned children, damaging their developing minds. Appalachian mountains were decapitated and stripped, ruining the land and water for farmers. For the first Earth Day on Wednesday, April 22, 1970, around twenty million people across the US participated in one of the tens of thousands of protests and activities to raise awareness around the environment. Energy use was one of the issues. Americans—at least, some Americans—were becoming aware of the daily trade-offs of consuming large amounts of coal, gas, and oil. In a prescient moment, on Tuesday, January 28, 1969, an oil well off the coast of Santa Barbara, California, burst and dumped thousands of barrels of petroleum into the ocean, creating an eight-hundred-square-mile oil slick in the water and a thirty-mile mess on the beachfront. Thousands of animals died. The accident captured how increasing demand for fuel required extracting energy resources with disturbing and deadly consequences.

So at the first public sessions over Zimmer, there was a small group of people present—the Fankhausers included—who expressed that CG&E should be investing in energy conservation, not more coal-, gas-, or uranium-fueled power plants. One woman asked, "What is going to happen to the atmosphere, to the water, to the land with the proliferating plants, not with just the Zimmer plant, but one plant right after the other?"[5]

This message hit a brick wall with CG&E. In 1972, as the Fankhausers were discussing conservation, CG&E executives reported to shareholders, "Sales of electricity continued their upward trend during 1972 . . . a gain of 6.1% over the previous year." They projected that over the next five years, electric sales would increase each year by a remarkable 7 percent. With that, they spent $114 million in 1972 alone on new property, plants, and equipment and planned similar expenditures for the future. These improvements were, in CG&E's words, "necessary to meet the increased demand for energy by existing customers and to meet the anticipated requirements for the future." Company leaders were operating in the same grow-and-build model that they always had and saw no room for cutting back on supply.[6]

Furthermore, CG&E officials bristled when residents lumped coal and gas with nuclear. Like other utilities investing in nuclear power stations, CG&E pitched uranium as an environmentally friendly, long-lived source of fuel. Of course, this wasn't quite accurate, as the Fankhausers and others pointed out at government hearings. The couple noted how Zimmer's physical parts and systems, like other commercial nuclear power plants, were expected to last around thirty-three years, not forever (although CG&E hoped the plant would survive longer than that; like other utility companies with nuclear power stations, CG&E banked on receiving at least one license renewal). After thirty-three years (or more, with a renewal), Zimmer would be decommissioned. Radioactive fuel rods would be removed and sent to a commercial low-level radioactive waste site; at Zimmer, leftover parts and structures would be entombed in concrete for a little over one hundred years.[7]

Prior to this final dismantlement, Zimmer was expected to use fifty-two thousand pounds of uranium per year. After individual fuel rods no longer yielded high levels of uranium for power production, they would have to be removed and buried at some distant waste site, and CG&E would have to install fresh ones inside the Zimmer reactor. For the spent rods, CG&E planned to send them to Barnwell, South Carolina, or West Valley, New York, where disposal facilities for low-level nuclear waste existed. (The material would travel in stainless steel casks on barges along the Ohio River, then by railroad.) The Fankhausers called attention to the long decay rate of uranium: even after rods were removed from Zimmer, they would continue to emit radiation for millions, if not billions, of years, depending on which uranium isotope was used.[8]

Furthermore, the federal government did not manage Barnwell and the other disposal facilities. Instead, beginning in the early 1960s, federal officials signed agreements with states for the burial of commercial low-level waste. Eager for economic development, states established burial sites that were then operated by private companies that used simple engineering designs to keep buried waste from infiltrating groundwater. This was before the National Environmental Policy Act of 1970 and before government officials and private contractors in the nuclear industry understood and acknowledged how radioactive material could migrate. Consequently, nuclear waste disposal companies in the 1960s did not systematically review the geological and hydrological features of a chosen area to determine the likelihood that buried waste would leak into the surrounding groundwater and soil. Instead, companies—with the support of state and federal officials, like those in the Atomic Energy Commission—brazenly cleared and graded land and then excavated shallow unlined trenches that were generally less than fifty feet deep to receive radioactive waste. After the waste was added, crews backfilled the trenches using recompacted material that had been removed during excavation. With these practices, unlined trenches often leaked radioactive material, which mixed with groundwater. This is what happened with the West Valley site. By 1976, it had leaked such high quantities of radioactive gas and liquid waste that its operator was forced to close the facility to deal with the contamination. The following year, the Maxey Flats Disposal Site in Kentucky, located 80 miles southeast of Zimmer, was ordered by the state to stop receiving radioactive waste due to extensive groundwater contamination. The site had received almost five million cubic feet of waste, including highly radioactive plutonium, uranium-233, and enriched uranium-235, which arrived in containers ranging from metal drums to cardboard boxes and was buried ten to thirty feet below ground.[9]

While the Fankhausers could not predict the closures of West Valley and Maxey Flats, they argued that in burial or reprocessing, nuclear fuel remained a potent, radioactive material—not a clean, renewable fuel. And while Zimmer's rods would be buried far away from Moscow, people from the area were still concerned about it. The principle of it bothered them. A local teacher from Moscow, Ohio, stated, "How can we justify our own increased use of electric power for air conditioning, night shopping centers lit up as bright as day, electric toothbrushes and gadgets if they mean we must leave such a heritage of accumulated radioactive waste to burden our descendants for millennia to come?"[10]

In 1962, Rachel Carson's *Silent Spring* pushed Americans to confront the effects of human action on the nonhuman world. She wrote about how the pesticide DDT, widely used for lawn maintenance, mutilated and killed fish and birds. People were beginning to realize the power of ecology—that everything is connected to everything, and we should take care with what we do to nature. As such, there were people at the 1972 Zimmer hearings who were concerned about the plant's impact on the natural world. Local environmental organizations—the antipollution Tri-State Air Committee and nature conservation groups like the Sierra Club, the Ohio Audubon Society, and the Ohio Izaak Walton League—took turns telling the Atomic Safety and Licensing Board that Zimmer's construction and operation would harm the surrounding area. A work crew had already done preliminary site work, razing eight farms, ten houses, and dense trees and vegetation on the property. Men had already excavated down fifty feet and then recompacted the soil to ensure a dense base for the plant, moving 22.5 million tons of earth in the process. Such uprooting and leveling work impacted nearby wildlife and contributed to soil erosion, causing surface runoff to flow into a nearby creek. Environmental groups were also worried about Zimmer's thermal pollution. Its discharged cooling waters would pour into the Ohio River, and since Zimmer's water would be significantly hotter than the Ohio's, there would be a loss of biodiversity from the heat difference.[11]

The Tri-State Air Committee and the others had gathered this information through public government reports. While the National Environmental Policy Act mandated environmental impact statements for federal projects, that didn't initially apply to nuclear power plants. It did after 1971, though, when a US District Court ordered the Atomic Energy Commission to comply. This was much to the chagrin of the commission, which wanted sole authority over nuclear matters. Regardless, the Atomic Energy Commission and a variety of federal and state agencies, along with CG&E, evaluated Zimmer's construction from 1970 to 1972.

Their reports troubled environmental organizations in Cincinnati. Aside from the ecological loss described in them by federal and state officials, people nervously noted how dismissive the Atomic Energy Commission and CG&E were. Despite the presence of deer, rabbits, red foxes, opossums, raccoons, and several species of birds in the area around Zimmer, commission staff cited "a paucity of flora and fauna present," so a reduction in population was "insignificant." Similarly, the commission trivialized Zimmer's water's impact on fish populations: "The size of thermally heated areas . . . is relatively small, at most a few acres. Although fish are attracted to areas of warmer waters at certain times, it is very doubtful that the numbers of fish attracted in this case would be significant," and "the fish could easily avoid these small unrestricted areas."[12]

Other agencies disagreed, though, showing the influence of environmental thought. Staff from the Department of Commerce wrote, "Although the dredging operation will destroy some organisms of no commercial value, these organisms may have important ecological value as integral parts of the food web. The intrinsic importance of these 'non-commercial' organisms should not be dismissed lightly, in the absence of sound evidence to the contrary." Representatives from the Secretary of the Interior's office and the EPA demanded more information on wildlife loss from chemical and thermal pollution and the impact of radioactive releases on nearby ecology during Zimmer's normal operations. Responding to this, CG&E representatives answered that the information already gathered was exhaustive enough and noted that some environmental impacts "were difficult to predict" but "will be insignificant." These pithy answers satisfied the Atomic Energy Commission, reinforcing people's impression that it was an apologist for utilities.[13]

Farmers and landowners from around Moscow were also worried about what radiation from Zimmer would do to nearby land and water. The Fankhausers' elderly neighbor, Elizabeth Seeburg, told the Atomic Safety and Licensing Board that she chose her home for its "lovely, beautiful surroundings" and that Zimmer threatened that. Born in Latvia in 1891, she came to Cincinnati to work as a clinical psychologist after earning her doctorate in psychology at the University of Wisconsin (one of the first women to do so). Seeburg purchased her property in New Richmond in 1950. Like the Fankhausers, she wanted to live as healthfully and naturally as possible and felt a nuclear power plant endangered her ability to do that: "I put in at least fifteen years of hard physical labor developing the land so I could have a fertile organic garden on which I could live in my old age and be useful in my old age. This plant is going to destroy that for me."[14]

Aside from these comments about conservation and ecology, it was human health and safety, not the land, that worried the majority of people at 1972 hearings. Parents were anxious about how radioactive isotopes could cause damage and mutations at the cellular level in their children, whose developing bodies were particularly sensitive. Over the 1950s, 1960s, and 1970s, fetal and infant health was on people's minds and in the news like never before. Health crises in these years sharpened parents' awareness of how vulnerable little developing bodies were—which, in turn, shaped the protest against Zimmer. The thalidomide disaster in the 1950s and early 1960s—where a drug given to pregnant women for nausea later caused serious birth deformities—terrifyingly put fetal and infant health in the spotlight. So did the aftermath of the drug diethylstilbestrol, given to women for the prevention of miscarriage: by the early 1970s, researchers had linked it to cancer in women and to developmental abnormalities and birth defects in children. Nuclear atmospheric testing that occurred around 1960 (and resulted in radioactive isotopes contaminating milk and wheat in the US) remained, a little over a decade later, a poignant specter for parents whose children would grow up near Zimmer. Additionally (and more positively), the ongoing reduction in infant mortality contributed to making people more vigilant about fetal health, and the ongoing women's rights movement helped women understand and celebrate bodies through pregnancy, childbirth, and early motherhood.[15]

Because of these developments, the generational effects of Zimmer took the spotlight at early hearings. "Our concern is not for ourselves," an older couple told the licensing board. "It is for what happens to the area and to the people that come after us." They likened nuclear power to DDT (the same pesticide Rachel Carson wrote about) in that DDT was used indiscriminately for years, despite health concerns, and only later declared carcinogenic. The couple stated that nuclear power, with its releases of radioactive isotopes, was similarly having a "cigarette moment" (initially thought to be fine, later discovered to be very dangerous). "Now, we know that many things have happened that have been okayed by the government," they said. "The use of DDT which had proved, for instance, to harm many of the species. . . . And now they are stopped. There have been drugs that have been on the market that have been approved by the government. And then all of a sudden we find these things are terrible, that they are doing things to the human being. Now, the same way with atomic energy."[16]

Parents James H. and Nancy Fichter, who had lived in New Richmond for fifteen years, were worried about their children. James, born in 1924, was from rural Butler County, Ohio, raised as a Jehovah's Witness, and Nancy, one year younger, was from the Columbus, Ohio, area. They met in Oxford, Ohio, where James's family lived and where he and Nancy were studying at Miami University. (James's father was a professor there.) While there, World War II broke out, and James, a medical student, claimed a draft exemption on the basis of his faith. When he failed to report for induction, he received a two-year sentence in federal prison. James subsequently enlisted in 1944. After his service, James and Nancy got married in 1948, moved to Oregon for a few years, then came back to Ohio, where James (who had switched career paths, from medicine to architecture) worked as an architect. In New Richmond, they had six children. "And one has already been born with a very bad kidney deformity," Nancy told the Atomic Safety and Licensing Board in 1972. The girl had been "over-exposed to X-ray through her illness. . . . We have done everything we could to keep her alive for the last eighteen years. . . . And we are terrified of this radiation that is coming forth from these plants and from the use of the atom bomb they have used in all of these years." Nancy continued, "Are we going to take this risk [of building Zimmer]? One man just said to me for progress. Do you call it progress? You should go to the hospital with my child and see if . . . you call that progress." Other parents brought their children with them to the hearings and even had them read statements, hoping to earn some sympathy this way.[17]

All of this wasn't just an expression of "not in my backyard!" (Usually shortened to NIMBYism and sounded out as *nim-bee-izz-em*, it refers to residents opposing development in their neighborhood, who are often satisfied if that development is just moved elsewhere.) Zimmer hearings instead revealed a collective, sophisticated idea of health and safety among Moscow-area residents. They understood that if an accident occurred at the plant, it would affect many people in the region. After all, only a few hundred people lived in Moscow, but twenty-five thousand lived within the surrounding ten miles, making Zimmer's public health considerations substantial. Elizabeth Seeburg, Fankhauser's neighbor, told the Atomic Safety and Licensing Board, "As a member of the human family, I am sufficiently identified with Clermont County [where Zimmer sat] to feel that things that go wrong with me go wrong with everybody else. That is my personal interest and it extends beyond. It extends to humanity." Furthermore, there were people who attended the hearings who didn't even live near Zimmer but instead resided in another part of the Cincinnati metropolitan area. When the board chairman asked one such woman if she lived "in the neighborhood," she replied, "Well, what do you call the neighborhood? I consider the neighborhood . . . this half of the state."[18]

And these protesters weren't just a group of back-to-the-land luddites opposed to new technology on principle. Hearing transcripts instead reveal middle-class, working-class, and farming families who saw a difference between sensible, appropriate technology and technology with risky—and possibly grotesque—consequences. Of course, as Zimmer's supporters would accuse them, most of these protesters didn't have a solution for area energy needs other than the insistence by some that the area didn't need additional energy development but rather conservation. Renewable energy like wind or solar—while known and discussed in the 1970s—would only begin to take off as an industry *after* the decade's energy crises. So as the Fankhausers, the Fichters, and others said no to nuclear power, that largely left coal for CG&E, which many in the immediate area around Zimmer opposed for its smog.

The protesters at the 1972 hearings are a good reminder of how much Zimmer is a story of evolution. The hundred or so residents who initially questioned the plant were a minority in their rural communities. Instead, many in and around Moscow really wanted Zimmer for its jobs and tax revenue. Fankhauser recollected many years later how some people in his community hated him for opposing the plant. He shared, "Despite being a Freedom Rider during the civil rights movement, it was Zimmer that found a burning cross in my yard." He was then and now under the impression that some of his neighbors did this.[19]

Outside of Moscow, Zimmer didn't engage very many people across the Cincinnati metropolitan region in the early 1970s. At the time, it likely seemed a remote threat (if it seemed like a threat at all). Since around 1970 there were so many pressing environmental matters, many Americans didn't see nuclear power as a burning issue, at least when compared to something as acute and awful as the Santa Barbara oil spill. Given this, early reactions to Zimmer were mimicked in other communities with proposed nuclear power plants. By the Shoreham nuclear power plant on Long Island in New York, the initial protest was mostly Long Islanders who lived nearby, as well as environmental groups. And in its early stages of permitting, the utility building the Seabrook nuclear power station in New Hampshire likewise encountered opposition only from local environmental and conservation groups. At the proposed Diablo Canyon nuclear power plant in California, it was the Sierra Club—not the masses—that initially fought the issue, although local community members, particularly mothers, soon joined the group. Much closer to Zimmer, the proposed Marble Hill nuclear power station in Madison, Indiana, similarly garnered the attention of concerned local environmentalists before anybody else raised a flag. To be fair, every protest movement starts somewhere, and many—including those that grow to make their mark on history—are initially limited in participants. After all, if we don't feel an issue is urgent, we tend to ignore it until there's a real crisis. Furthermore, nuclear power was a unique environmental issue in that it divided even environmentalists. While some were troubled by its radiation emissions, others concluded that it *was* a cleaner fuel compared to coal. Cincinnati's Air Pollution Control League backed nuclear power for this reason, and even the national Sierra Club didn't become critical of it until 1972.[20]

Nevertheless, while the protest against Zimmer was small in 1972, the fact that a few hundred people showed up to permit hearings still aggravated CG&E officials. The utility's representatives and legal counsel tried to placate people's fears by stressing the company's competency to build Zimmer and reminded people that nuclear power was highly technical, easily misunderstood by untrained persons. This latter comment was directed at people like Nancy and James Fichter. They weren't like David Fankhauser with his PhD in biology. But the Fichters *were* citizens, residents and homeowners in the area, and parents, all of which in their minds qualified them to have and voice an opinion and to be taken seriously. They insisted on their right to be heard. The result was a major clash with government officials, utility representatives, and their supporters.[21]

Prior to the first licensing board hearing, the Fichters applied for intervenor status, as did David Fankhauser and Elizabeth Seeburg. Intervention, also called *legal standing*, gave a petitioner the right to participate fully in proceedings. Without standing, a member of the public would have their statement inserted into the record, but it was not considered official evidence, diminishing its power. So intervention was very powerful, but people only achieved it when they demonstrated a sufficient interest in the matter at hand.

Standing had long been used in citizen battles against environmental abuse by corporations, but during the nineteenth and early twentieth centuries, courts granted intervention only to petitioners who proved that they or their property were being financially harmed. Beginning in the 1960s, thanks to the growing environmental movement, courts began to broaden their understanding of what constituted an "interest." The preservation battle over Storm King Mountain in the late 1960s was key to this. New York City's utility, Con Edison, proposed a hydroelectric power plant at the base of the scenic Storm Mountain in the Hudson Highlands. After the Federal Power Commission granted Con Edison a license for the plant, environmentalists sued the commission. While Con Edison argued in an appeal that environmentalists had no standing to sue, the court of appeals agreed with the environmentalists, giving them standing to sue a federal agency in court. Thereafter, federal agencies issuing licenses with environmental implications began to open their proceedings to allow for additional intervenors with an environmental interest. For that reason, Zimmer wasn't the only nuclear power station that saw different citizens, environmental groups, and others taking advantage of this opening and applying for intervention in permit hearings.[22]

Fankhauser and the others based their requests for standing on a variety of factors. They lived close to Zimmer. They had children. They believed in energy conservation. They thought other technology could power the region. They had serious reservations about nuclear power's safety. It was a mixture of ethical, scientific, and health reasons, but after reading their requests for intervention, the Atomic Safety and Licensing Board denied all petitioners.

The problem was that Fankhauser's interest, like the others petitioners', was too general. CG&E legal counsel noted at the time that "all of the contentions by the petitioners basically attack the general policies of the United States as established by the Atomic Energy Act . . . this proceeding is not the proper place for such challenges." Atomic Safety and Licensing Board chairman Farmakides echoed this, saying, "Dr. Fankhauser, similarly, made very general assertions relating to all nuclear plants. . . . While this Board does not question the sincerity of these persons, it cannot permit their intervention based on such vague and unparticularized assertions." This pleased CG&E leadership. CG&E executive William Dickhoner commented at the time, "We are gratified that the hearing will proceed without intervenors who might have delayed the project and increased the cost."[23]

The rejections showed that even as intervention had expanded over the 1960s to be an invaluable tool for activists, it was still not a guaranteed legal strategy. Denied standing, Fankhauser and the others swallowed their rejections and fought to use their allotted time (about five minutes each) as best as possible, inserting into proceedings their broader critiques of nuclear power. Fankhauser stated to the board at one hearing, "Now, I am a molecular geneticist. And I did considerable research on mutations and induction of mutations at Johns Hopkins University. And part of my research included the induction of mutations by an ionizing radiation. And I would like to . . . bring before the Commission and impress upon the Commission and the public the severe dangers involved in this radiation. . . . And I would like to furthermore question not only the safety of this particular station, but I would like to question the safety of any station."

The licensing board quickly reminded him that such broad critiques of nuclear power would not be tolerated—or as CG&E legal counsel put it, "You are in the wrong courthouse." Those in opposition to Zimmer also tried to lengthen their allocated time. As Farmakides opened the first hearing on Zimmer, Nancy Fichter interrupted him, asking, "May we have a better explanation as to why we only get five minutes? That doesn't seem very important." Farmakides replied, "This is a ruling of the Board. So that is finished insofar as I am concerned."[24]

Her interest in gaining intervenor status—and challenging a man like Farmakides—reflected a new culture of citizen participation in policy by the 1960s. Thanks to court cases like Storm King Mountain and to community organizing by the civil rights movement, several federal environmental policy acts—including the Clean Air Act of 1963, the Air Quality Act of 1967, and the National Environmental Policy Act in 1970—encouraged, if not required, public involvement (like public hearings) in decision-making. People like Fankhauser and Fichter came to expect by 1972 that their voices should be heard. "Now, I know many of the officials of [Cincinnati] Gas and Electric," one woman stated to the board. "And they are fine gentlemen. They are working for their plant. And they believe naturally—they are interested in their interest in their plant and naturally they are going to take their plant's view of it. And I cannot blame them for that. But we ought to be told some things too. We ought to know what is going on."[25]

Her expectations clashed with those of board members—like the short-tempered Farmakides—and this same script played out at other contested nuclear power plants, where members of the public felt they weren't welcome or heard by Atomic Energy Commission and Atomic Safety and Licensing Board members. Licensing board members were often men like Farmakides, trained and immersed in bureaucratic and military cultures. Farmakides was a Greek immigrant who, in the early 1970s, was only in his forties but nonetheless had an impressive government and military resume. After naturalization, he served as a colonel in the US Army in the Korean War and later worked as a judge and attorney in several government positions, including the Atomic Energy Commission, the National Science Foundation, NASA, and later the US Department of Energy. And while Atomic Safety and Licensing Board members were often like Farmakides in that they had worked for the AEC or an affiliated laboratory, they were subject to strict conflict-of-interest laws that forbade them from being personally and financially involved with any nuclear plant or project. Viewing themselves as objective reviewers, they aimed to get nuclear plants operable as expediently as possible and had little patience for commentators like Fankhauser who slowed the already lengthy review process. Furthermore, board members restricted discussion at hearings to specific topics, determined to prevent their boards from becoming debate forums over national nuclear industry standards. This was a major point of contention with members of the public who wanted to discuss a broader scope of information, such as the basic question of whether nuclear power was even a good idea. Furthermore, atomic licensing boards were composed of one administrative law expert and two technical experts. (One of the people serving with Farmakides, for instance, worked at the Accelerator-Based Conversion Laboratory at Los Alamos, New Mexico, where the federal government manufactured nuclear weapons.) These were highly trained individuals who for the most part believed in the scientific claims made by the Atomic Energy Commission. Given this, they were disgruntled when (less educated) members of the public made statements about nuclear technology.[26]

People also—and especially—clashed with CG&E's sharp-tongued attorney Troy B. Connor Jr., who, unlike members of the public, could—and did—speak at length in the hearings. Hired from Washington, DC, Connor had provided counsel for other utilities trying to get nuclear power plants to licensure. (He was known as *that* lawyer you hired to get the job done.) At Zimmer's hearings, he was intent on showing the Atomic Safety and Licensing Board how popular and needed Zimmer was. He brought in supportive local and state officials who were eager to see Zimmer provide high-capacity power, and he skillfully solicited appearances from pronuclear engineering experts, contrasting their opinions with those of someone like Nancy Fichter.

James H. Leonard, a University of Cincinnati professor and the consulting engineer for Zimmer, was one of them. Head of chemical engineering at Cincinnati from 1967 to 1973, Leonard was also the director of the university's nuclear engineering program starting in 1966, when he was hired. Prior to that, he worked as a specialist in reactor design and control at a laboratory affiliated with the Atomic Energy Commission and helped to design the nation's first commercial nuclear reactor, located in Shippingport, Pennsylvania. Leonard's colleague at the university, James N. Anno, another nuclear engineering expert, with service in both commercial and higher education aspects of the technology, also served as a CG&E ally during Atomic Safety and Licensing Board hearings. The two men were like many other nuclear scientists who had worked in government-affiliated jobs and contracts: from such experiences, they had a vested interest in supporting Zimmer—and nuclear power in general—and genuinely believed in the technology's merit.

At public hearings, Leonard and Anno explained that people could trust the engineers and contractors designing and building Zimmer. They also assured people that the AEC knew how to regulate the construction and operation of nuclear power stations. Anno told the licensing board at Zimmer's June 1, 1972, hearing, "In my opinion, the Zimmer plant is an entirely acceptable project. I believe the proposed plant to be safe and not to present undue hazards to the public or the environment."[27]

But as soon as he sat down, a member of the public in the back of the room shouted (out of turn), "Would it be appropriate to know what biases this man might have who presented this statement? . . . Is he presently receiving grants from the Atomic Energy Commission or the Cincinnati Gas & Electric?" Farmakides silenced the question with a simple "no." Indeed, Zimmer's early skeptics didn't miss how Farmakides and his board treated them differently from Leonard and Anno. Lawrence W. Kessler—the director of the local environmental group, the Tri-State Air Committee, as well as a University of Cincinnati law professor—stood up at one public hearing to condemn the "self-imposed isolation by this [licensing board] to insulate its decision-making process from the effective presentation of unfriendly views." He later decided to pull his organization from attending future Zimmer hearings out of exasperation. Denouncing the "fraud" and "sham" of the hearings—his exact words—he vented that CG&E and government reviewers had sidestepped his questions about the plant's safety systems one too many times. "Everything was done to protect the applicants and the commission from critical examination," Kessler stated. Another man named Stan Hedeen agreed. He was a biology professor at Xavier University in Cincinnati and the conservation chief for the Ohio chapter of the Sierra Club. He told Farmakides, "If the purpose of this hearing is to determine if the Zimmer plant can be built without risk to the public health and safety, then why did the AEC allow construction to begin before this hearing?" The audience shouted in agreement. "I would agree with the professional engineers," Hedeen continued, "that they feel these hearings are a farce." He then admonished CG&E to rewrite its construction permit, using past tense.[28]

Hedeen and other scientists opposed to Zimmer used their precious minutes as best as they could, using the time to both argue with CG&E and educate the public about nuclear power. In this way, they acted like other contemporary scientists like Rachel Carson or Barry Commoner, a nationally known biologist, nuclear power critic, and tireless advocate for environmental health and safety. These activist-scientists, who grew in number in the 1970s, believed they had a duty to disseminate public health research to people in easy-to-understand ways, and in doing so, they not only advanced environmental issues, but they also disrupted the prestige that the Atomic Energy Commission had by showing that governmental officials and contractors weren't the only experts on nuclear power.[29]

At one hearing, Fankhauser explained what an emergency core cooling system within a nuclear reactor was. "I am sure that the members of the Board are thoroughly familiar with this and the public, I think, probably is not," he said. "The Emergency Core Cooling System, which is often abbreviated ECCS, is a system that in the case a reactor—" The licensing board cut him off and told him to stop trying to educate people. They said his concerns over ECCS—that, in Fankhauser's words, "a meltdown accident would release phenomenal amounts of radioactivity into the river"—were not specific enough and thus irrelevant to the Zimmer hearing.[30]

People at the hearings were not deterred, though. Even those who did not hold advanced technical degrees were avid readers and followers of the latest scientific research on radiation's health impacts. Referencing critical studies and articles in their statements to the licensing board, they stressed that nuclear power's risks outweighed its benefits. They questioned the prudence of a technology that *could be* very dangerous. Not afraid to cut in and cut off government reviewers, these early protesters at Zimmer forcibly made the hearings more democratic.

CG&E leaders were very confident in the engineering of nuclear reactors, so they downplayed the possibility of an accident at Zimmer, where a significant dose of high-level radiation could be released. At this time, nuclear engineers built reactors by imagining worst-case, far-fetched catastrophes and developing conservative designs on that basis, using multiple, redundant, self-correcting safety features to prevent or mitigate extreme accidents. These features gave rise to the term *fail-safe*, where nuclear power advocates believed that reactors' safety systems would always prevent or mitigate serious accidents. CG&E leaders said as much: "In the first place," one of its executives noted, "the design, materials, workmanship, and quality assurance programs for the plant offer great assurance that this incident [a serious accident] will not occur." At early public hearings, CG&E's lawyer defended Zimmer's safety by pointing to its various "fail-safe" aspects. He noted that in the reactor core, cruciform-shaped control rods—containing neutron-absorbing elements like boron or cadmium—would move vertically in between the assembles of uranium fuel rods, in the process absorbing neutrons to either slow down or speed up the reaction. Safety relief valves, located throughout the pipes carrying steam from the reactor to the turbine, would discharge steam if pressure extended beyond a certain point. Additionally, Zimmer, like other nuclear power plants, would have an emergency core cooling system (an ECCS, the system Fankhauser had tried to explain at a hearing), where the reactor core would be flooded with water if its coolant malfunctioned. If the reactor still overheated, additional barriers—including the containment building and the pressure vessel surrounding the reactor—would trap the release of the fission products in the core. Zimmer's safety was contingent on these physical systems preventing excess radiation from escaping. Since there were thousands of safety components within one nuclear power plant, nuclear engineers in the 1960s—when so many utilities were ordering nuclear power stations—did not invest in determining and limiting the known probabilities of accidents.[31]

The logic of—and confidence behind—the fail-safe systems did not reassure people at Zimmer's first hearings. One member of the Ohio Audubon Society, who was also a chemist and physicist with a PhD in high temperature and energy conversion, was convinced that existing design protocol was not enough to guarantee *total* safety—that CG&E had not planned for the "the maximum hypothetical accident." To this point, he posed the larger issue of: What was the *maximum* possible accident? Were the engineers—he asked—who were designing nuclear reactors based on the worst events that could credibly happen considering truly *incredible* events? CG&E officials and board members expressed that this line of questioning was hyperbolic, but not everyone in the room agreed with them: by the 1960s and 1970s, Americans were interested in living more risk-averse than ever before. More people were wearing seat belts in their cars; more people were quitting smoking cigarettes. The risk of a major nuclear power plant accident—which could be acute and long-lasting in health impacts and possibly grotesque in its toll on human bodies—was scary.[32]

The Atomic Energy Commission itself increased people's fear of an accident. As more utilities ordered nuclear power plants, and as communities underwent public hearings over them, it struck a growing number of people that the commission was too promotional in its work, leading to the possibility of unsafe nuclear power plants. Furthermore, the commission relied on a program of quality assurance (often abbreviated as QA) to verify that the fail-safe systems had been installed correctly at nuclear power plants. Quality assurance included all the systematized actions, inspections, recordkeeping, and audits that went into a nuclear power plant. Importantly, the commission largely outsourced this work to the utilities and contractors building the stations, which CG&E and other utility executives appreciated.[33]

Fankhauser and others at Zimmer's hearings took issue with this. They also pointed out that emergency core cooling systems were undergoing national public hearings over disputed efficacy. Many asked: Shouldn't we pause on Zimmer since the plant had such systems? The Atomic Safety and Licensing Board dismissed the request. Still, the dissension over key fail-safe systems revealed that the AEC was having trouble with its regulatory load. It was short-staffed, and the number of plants it was reviewing was growing unsustainably. More plants were being ordered, and they were getting bigger and more varied in design. By the early 1970s, AEC staff started to design new (post-Zimmer) reactors using more sophisticated quantitative risk assessments with early computer modeling. One thing didn't change, though: AEC officials' belief that nuclear power was safe. One report it issued, which used its new methodology, said that risk of death from a nuclear power accident was comparable to being struck by a meteorite.[34]

If government and CG&E officials had a hard time understanding people's concern over an accident at Zimmer, they certainly did not understand people's concern over Zimmer during its *normal* operations. At hearings, they told members of the public not to worry about the low-level radioactive gaseous and liquid discharges that would come from Zimmer as it operated. CG&E executive William Dickhoner promised, "We are going to design this plant so the maximum (radiation emitted) will be one per cent of the standard (imposed by the AEC)." Dickhoner and his colleagues believed, as did other advocates of nuclear energy, that under a certain amount of millirems, the body was safe (a millirem is a measurement of radiation, one rem being a large dose). Because the utility's leaders held faith in government standards, and because Zimmer's emissions were based on AEC standards, they believed Zimmer's risk was insignificant. At hearings, Atomic Safety and Licensing Board members supported this view. One of them declared that the safety record of the nuclear industry was "almost impeccable." "The big explosion has not occurred," he continued. "Fatalities have not been traced to radioactive wastes. We live in a modern world, populated by modern people with highly sophisticated scientists and physicists. They have and are perfecting wonderful things every day to make our lives more comfortable and healthful."[35]

But Fankhauser and others questioned that logic. Even below a certain threshold, they said, low-level radiation was not innocuous. "No self-respecting geneticist would buy [that]," Fankhauser told the licensing board. He argued that nuclear power plants, in daily operations, increased background radiation to dangerous levels. "Man can handle eighty millirads a year. We know that. Because he evolved in that level. There is no question that if you increase this to eighty-one [millirads] that the mutations induced will be increased proportionately to this one millirad." More to the point, he stated, "Every single dose of radiation carries a finite risk." Continuing, he told the board:

> This radiation . . . severely endangers myself, my wife, my child, my neighbors, and, more importantly, which I hope everybody can understand, it is not just me, it is not just my children. It is my children's children that once DNA is damaged it is not repaired. The effects of low-level radiation may not be seen for several generations. And, of course, by that time people then will look back to these hearings and say, "Why did the public allow this to happen?" And they will then also, I suspect, look at the way that the government is set up and the way that this Board is set up, such that it prevents the public from having a say and from having a complete say.[36]

Fankhauser's caution over low-level radiation echoed ongoing scientific arguments over how much, if any, radiation could be emitted safely from nuclear power plants. It had long been established that radioactive isotopes were linked to cancer and that the longer the half-life of a radioactive isotope, the worse the damage was once it entered an organism and accumulated in sensitive organs. Since studies showed the higher the dose an animal received, the higher the chance for a mutation, some scientists argued that below a certain threshold, there was a tolerable dose of radiation that would not cause damage. This was the official position of the Atomic Energy Commission and its affiliated scientists—at least until the 1960s, when a trickle of researchers began to dissent and argue that government standards for permissible low-level doses of radiation were too low. Some even concluded that thousands of additional cancer cases would arise annually if a population received the AEC's allowable number of rads (absorbed radiation) from operable nuclear power plants.[37]

But the issue was how anybody could quantify—with precision—the actual risks of low-level, casual exposure from operable nuclear power plants. Scientists experimenting on animals understood that radioactive isotopes affected no two bodies the same way, and they also knew that low-dosage radiation had a latent impact on the body. That made it difficult to conclude that someone's exposure to radioactivity was *the* or *a* cause of a person's (later) illness. Furthermore, to understand low-level radiation's long-term effects, scientists extrapolated conclusions from high-level emissions, using data from studies on Japanese bomb survivors from World War II. But no one knew if cell mutations would be the same—just slower—in response to low-level radiation as they would to a large release of radiation.

In 1970, the Atomic Energy Commission published its new guidelines for nuclear power plant emissions. It dictated that utilities should keep radioactive emissions "as low as practicable." The next year, the commission set more concrete standards, providing a few ways that nuclear power plants could gain licensure. For instance, if emissions from the plant were less than 5 percent of natural background radiation (or about 1 percent of the maximum recommended exposure for an individual member of the public), then the plant met AEC standards. Or if an individual living at the perimeter of the plant would not receive more than five millirems to the skin per year, then the plant met AEC standards. Indicative that its primary goal was the promotion and not regulation of nuclear power, the AEC included an "operating flexibility" in its revised standards so that plants could stay open if their releases temporarily breached upper limits. Throughout the 1970s, radiation standards continued to change in response to heightened scrutiny, growing people's unease.[38]

These debates influenced Zimmer's early hearings: worried residents and civic groups noted the vacillating standards. Nancy Fichter asked the licensing board at the first Zimmer hearing, "Is it true that they are lowering the levels that they are allowing? . . . Well, don't you feel that it is significant that it is getting lower and lower and more people are questioning this, more scientists are questioning this?" At the next public meeting, she stated, "What permissible actually means is that only so many are going to die." Lawrence W. Kessler—the director of the Tri-State

Air Committee—urged the licensing board to take seriously the health risks of low-level radiation, for it could turn out that its effects were more damaging than previously thought. Air pollution had had a similar trajectory, after all. Kessler stated, "We are in a comparable position in our understanding of radiation emissions to what we were 30 years ago with regard to sulfur oxide, carbon oxide and other emissions from standard generating facilities. Those were built without concern for the possible health damage. We now suffer an estimated several thousand deaths a year directly attributable to air pollution. There is no reason to make the same mistake again."[39]

The story of Zimmer, at this stage at least, is not about determining how safe nuclear power was. That was a matter of national and expert debate. Rather, Zimmer shows how that scientific debate was impacting ordinary people. That dissension planted a seed in people's minds that nuclear energy, in its functioning and in its potential to malfunction, contained a risk. Even nuclear power's most ardent advocates conceded that. It was, then, a matter of "how much risk is it?" and "what do we do with that?" A small but growing number of Americans were concluding that they'd rather not take the risk at all, especially when their health, and that of their children, was at stake—that it was better to be cautious since nuclear power, as a manmade technology, would probably have some error. People formed a hard line on this matter, and pronuclear experts, convinced that the risk wasn't significant, took a different hard line—hence, tense public hearings where people struggled to talk over these lines.

In the summer of 1972, the sessions over Zimmer's construction concluded. The Atomic Safety and Licensing Board had listened to Fankhauser and other concerned residents—but that was it, it seemed. That fall, the Atomic Energy Commission finished its environmental review of Zimmer's construction, assessing that the building of the plant met AEC standards as well as those of other government agencies. The Advisory Committee on Reactor

Safeguards also gave Zimmer its go-ahead for its construction. On Monday, October 16, 1972, the AEC granted CG&E a construction permit, and in the spring of 1975, CG&E officially applied for an operational permit. The 1954 Atomic Energy Act laid out a two-year review process for utilities to get operational licenses, during which regulatory bodies conducted reviews and wrote up reports on the plant, much like they did for the construction permit. These reports triggered another round of public hearings, where a new Atomic Safety and Licensing Board would convene to discuss the findings. If satisfied, AEC staff would then issue the operating license.

In 1972, CG&E expected Zimmer to be operable by 1977. That did not happen. Instead, it took the Atomic Energy Commission and other government bodies until 1979 to finish reviewing Zimmer. Then, with the reports available, public hearings commenced (again) in 1979, initiating another lengthy portion of the licensure process. Certain features of Zimmer—its General Electric Mark II containment vessel around the reactor, only one of seventeen under construction or in operation across the world—meant heightened scrutiny and slowed the review process. But nuclear plant licensing in the 1960s and 1970s was just generally sluggish due to regulatory bodies' limited staff and the growing number of power plants needing review. In the early 1970s, different directors of the Atomic Energy Commission increased its size and budget and urged speed with license review, hoping to improve the commission's efficiency and public reputation—but these efforts were to little avail. Then, with the support of much of the AEC's staff, President Nixon, followed by Gerald Ford, shepherded legislation through Congress to dissolve the Atomic Energy Commission and divide its functions into a new regulating/licensing Nuclear Regulatory Commission (the NRC) and a nonregulatory Energy Research and Development Agency (which in 1977 was combined with the Federal Energy Administration to form the Department of Energy). And yet, even with these dramatic changes, it still took years to get a plant into operation. In 1967, the average time to plan, build, and initiate a nuclear power plant was seven years. By 1980, it took on average ten to twelve years.[40]

During the long delay between the issuance of Zimmer's construction permit in 1972 and the reopening of licensing hearings in 1979, CG&E made significant headway on Zimmer's construction—in fact, the plant was over 90 percent complete by 1979. Still, despite such an obvious sign that Zimmer would become functional, protest continued. In fact, Fankhauser, the city of Cincinnati, and a new consumer rights group, the Miami Valley Power Project, earned legal standing in 1976 in an effort to thwart CG&E's operational license for Zimmer. Miami Valley Power Project had around one hundred members (many of them lawyers). It was formed in Dayton, Ohio, in 1974 to protect Dayton Power and Light ratepayers (one of the co-owning utilities of Zimmer) if the nuclear power plant did not open or operate efficiently. That very mission was Miami Valley's justification for intervention—which worked, for it was a very "specific" interest. Fankhauser, too, made his reasons for legal standing more specific to Zimmer. He hired a helpful lawyer from his town of New Richmond, and together they focused his application on the fact that he and his family lived near Zimmer and were threatened by its potential health effects.[41]

The city of Cincinnati petitioned for intervention on the basis that Zimmer sat nineteen miles upstream from city water intakes, making its discharged waters a matter of relevance. Put more bluntly and less diplomatically, city officials did not trust CG&E to adequately monitor those discharges.

There was a backstory to this. After Nixon created the EPA in 1970, the EPA and the Atomic Energy Commission fought for control over who should regulate emissions from nuclear power plants. With the 1972 Clean Water Act, the EPA made it unlawful for anyone to discharge a pollutant from a point (singular) source into navigable waters without a permit. This then raised the question of whether the EPA should have any role in supervising water discharges from nuclear power stations. In 1973, the Office of Management and Budget gave the AEC that role while mandating that the EPA monitor other radioactivity. The AEC, followed by its successor, the Nuclear Regulatory Commission, did not perform this work itself, though. Rather, it devolved monitoring responsibilities to utilities and other operators of nuclear facilities. As such, CG&E said it would monitor Zimmer "as comprehensively as possible." The ambiguity worried city of Cincinnati officials, who instead wanted independent, extensive, and continual monitoring of water outtakes at Zimmer. (CG&E stated in its environmental report on Zimmer that after the first three years of monitoring water discharges, the monitoring program would be reduced if radiation doses were shown to be "sufficiently small.") The city wanted data on water to be directly transmitted to its Water Works; officials also asked for CG&E to include the city in its emergency planning and for CG&E's environmental review board to include a city of Cincinnati representative. City officials concluded that the only way to secure these outcomes over water safety was to intervene in Zimmer's licensing hearings.[42]

In January 1976, city of Cincinnati assistant solicitor William Peter Heile said as much in front of a new Atomic Safety and Licensing Board, and its chairman, a man named Samuel Jensch, agreed that the issues over city drinking water constituted a "specific interest." In this way, Jensch was significant: he demonstrated that not everyone connected to the Atomic Energy Commission and its licensing boards was standoffish to the public. Rather, some officials *were* interested in feedback (and, in fact, over the 1970s, more officials connected to the AEC would become vocally critical of its insularity). Jensch—while still being "an AEC man" who believed in nuclear power—nonetheless helped Zimmer's petitioners gain standing by being flexible and open-minded.

Born in 1908 in beautiful Hudson, Wisconsin, near Minneapolis, Jensch served in the US Coast Guard during World War II and then had a long legal career with the federal government, including being an administrative law judge for the AEC. His high school yearbooks and World War II draft card show him as a man of medium height with a long, straight nose, black hair, gray eyes, and many freckles. He and his wife, Beverly, with whom he had two daughters, were nature lovers. They had to live in Washington, DC, for Sam's work, but spare time saw them back in rural upper Wisconsin. Beverly was also a historic preservationist, an active member of the Congress of Racial Equality (the same civil rights organization that Fankhauser was involved in), and a women's rights advocate. (Perhaps she influenced some of her husband's open-mindedness.)[43]

Returning to January 1976, when city of Cincinnati solicitor William Heile expressed concern over discharges from Zimmer, Jensch showed sympathy with Heile's points, mainly by arguing that there was still ambiguity in environmental law and jurisdiction around who should monitor radioactive (and otherwise hazardous) discharges, and given that, Heile's contentions were valid petitionings. Jensch, for instance, brought up a 1975 court case, *Colorado Public Interest Research Group vs. Train* (*Train* for Russell Train, the then-EPA administrator). Jensch said, "I understand there is a case, I believe it has been accepted by the Supreme Court, involving Colorado water . . . EPA has been held by a lower court to have the responsibility of determining I guess radioactivity in the water, and there is a contention that NRC should have that jurisdiction." The EPA, according to its agreement with the Atomic Energy Commission and then the Nuclear Regulatory Commission, disclaimed authority under the Clean Water Act of radioactive discharges. But in 1975, residents and organizations near the Fort St. Vrain Nuclear Generating Station and the Rocky Flats nuclear weapons plant in Colorado sued the EPA over the plants' effluents (wastes), arguing that radioactive waste constituted a pollutant. In short, they wanted the EPA, not the Nuclear Regulatory Commission, to monitor the plants' discharges. The Clean Water Act, like other federal legislation of the time, permitted "citizen suits" when citizens felt the EPA had failed to perform nondiscretionary duties. Heile argued (and Jensch agreed) that this court case was proof that different communities—not just Cincinnati—saw the need for extensive and independent monitoring of nuclear technology. (In June 1976, a Court of Appeals upheld that the EPA should monitor all radioactive pollutants going into water, even if they were also regulated by the NRC.)[44]

Heile also wanted CG&E to pay for and monitor Zimmer's discharges "at the source," insisting that the city—being downstream—should not have to clean up water with trace amounts of radioactivity. "I think it is well-respected, anti-pollution practice to eliminate pollution at its source rather than through the purification method," Heile explained. Jensch concurred, pointing to the example of the Reserve Mining Company near Duluth, Minnesota. Beginning in the 1950s, it dumped thousands of tons of taconite tailings into Lake Superior, contaminating Duluth's water supply (tailings are the waste byproduct of extracting iron ore from taconite rock). In 1973, the EPA sued the mining company, and the following year, it shut down the operation. The company appealed the decision and reopened, disposing its waste on land instead of Lake Superior. Jensch said it was a useful ongoing legal battle over if and how private companies had to tackle pollution "at the source."[45]

After talking through these examples, Jensch told the city solicitor that these environmental battles showed how regulations "are on the moving target basis; things change as these things occur." Furthermore, Jensch said that until a nuclear power plant was actually constructed, "you really don't know what it's going to be like." Acknowledging that, he believed, "would open up a little inquiry"—meaning that topics discussed at early Zimmer hearings might later need an updated review. For instance, in 1972, the Federal Power Commission concluded that "electric power output of the Zimmer Unit No. 1 is needed to meet the Applicant's projected loads." Four years later, at the hearing with Jensch, the Miami Valley Power Project—thinking about Dayton, Ohio, ratepayers—claimed that those projections were no longer valid—that they had been overestimated. When CG&E representatives insisted that demand projections were still good, Jensch fired back, "You think they are cast in cement at the construction permit stage?"[46]

Jensch further astounded CG&E officials when he agreed with Cincinnati's solicitor that a private citizen should sit on CG&E's environmental review board for Zimmer. Jensch said that having a "good listening post" and "an open door on policy data" could help CG&E with its community relations. And when the utility continued to insist that residents and groups should not be given intervenor status until their applications were further "specified," Jensch attacked this vague requirement, calling it "nitpicking." "You can always conceive of some indefinite definition—if it only had said Tuesday instead of Wednesday," he said. CG&E legal counsel responded, "If they [those applying for intervention] were to present expert opinion on an expert basis—" Jensch cut them off, asking, "Who's an expert?" He pointed out that the CG&E lawyer, unlike the petitioners, was not from the region. His point: that residents of an area are its best experts.[47]

Ultimately, Jensch and his board recommended legal standing for Fankhauser, the city of Cincinnati, and the Miami Valley Power Project, and the Nuclear Regulatory Commission granted it in March 1976. Over the course of the next several years, the right to officially intervene proved to be a double-edged sword for Fankhauser and the others. It ate up years of their lives. Prior to each hearing, intervenors had to answer queries from the utility and read through hundreds of pages of legalese and technical information about nuclear engineering. This, in Fankhauser's mind, seemed intentionally designed to embarrass and overwhelm him. Quickly, he amassed boxes and binders full of thousands of technical documents. Two such reports were "Cladding, Swelling, and Rupture Models for LOCA [loss of coolant accident] Analysis" and "Anticipated Transients without Scram for Light Water Reactors."[48]

Still, from the activists' perspectives, the positives of legal standing outweighed its demands. As most of the fight against Zimmer took place in hearing rooms, intervenors could agitate against Zimmer by forcing government reviewers to discuss their contentions. It enabled them to participate and submit "interrogatories" to CG&E, which were considered official evidence and could sway licensing board decisions one way or another. When the city first received standing, Heile and another city solicitor submitted to CG&E an exhaustive fifty-five pages of questions, trying to determine exactly how the utility would monitor discharges. This early involvement of the city of Cincinnati in Zimmer's licensure foreshadowed its later, and larger, role in fighting to make sure Zimmer was built safely and fairly. Despite the fact that Zimmer wasn't even in the city's boundaries, municipal officials—different mayors, city council members, and city staff—took seriously environmental health threats and financial risks to the community members they served. By questioning CG&E's project, city leaders—in charge of around 400,000 incorporated residents in the 1970s—encouraged others in the area to do so too.[49]

3

"The Public Has to Make the Decisions"

The City and Zimmer

"There can be no doubt that if any radioactive substances are released into the Ohio River these will find their way into the drinking water of Cincinnati residents." These were the words of a young Cincinnati city councilor, Jerry Springer, at a Holiday Inn north of Cincinnati on Thursday, June 1, 1972. There, in the hotel's small banquet hall, a room less than six hundred square feet large, close to one hundred people had packed themselves in. This was one of the early Atomic Safety and Licensing Board hearings, convened before CG&E had been granted a construction permit.

Springer had been recently elected on the Democratic ticket, and his words marked the first time a city official said anything critical of Zimmer. He explained his reservations about the plant—that, once operable, Zimmer would discharge slightly radioactive water into the Ohio River, where Cincinnati sourced most of its drinking water. Springer told the Atomic Safety and Licensing Board, "The Atomic Energy Commission alleges that the water radiation will not be harmful. Yet, in making this allegation, they rely upon a still unproven theory that the body can repair damage caused by low level radiation. There is contradictory evidence which points to an increase in instances of cancer, leukemia and genetic mutation caused by an increase in the level of radiation. The best that can be said about this conflict of authority is that the effect of long-term low radiation on the human body is still largely unknown."

He urged that, until "there exists no medical doubt as to the effects of low-level radiation on human beings and the environment," the licensing board should pause Zimmer's licensure process. Springer likened the Atomic Energy Commission to the "industrialist of thirty years ago who might have responded to complaints about black smoke billowing from his smokestacks saying 'So what, nobody has proven that smoke is bad.'" Applause broke out. David Fankhauser was one of those clapping.[1]

Springer was just getting started. In fact, he—and a majority of his fellow city councilors, plus city staff and multiple mayors—would spend the 1970s challenging Zimmer. A few developments propelled this course of action. In 1972, the city convened its first-ever environmental task force, comprising volunteer citizens and local experts who cared about various environmental issues confronting the city. Its energy subcommittee passionately debated Zimmer and the extent to which its discharges could impact the city's drinking water. As it was doing so, University of Pittsburgh radiation physicist Ernest Sternglass published a paper claiming that people along the Ohio River Valley were dying from heightened rates of cancer because of the nuclear reactor in Shippingport, Pennsylvania. The reactor sat upstream from Cincinnati and was discharging effluent into the Ohio River.

The Sternglass report terrified those who believed it—and angered those who took issue with its conclusions. Still, even as local, state, and federal officials hotly debated the paper, and even though it was about Shippingport and not Zimmer, the city's environmental task force ultimately recommended that "the City of Cincinnati should exert every effort to forestall operation of the Zimmer Power Station," despite the fact that Zimmer wasn't in the city's jurisdiction. Springer and other council members listened.

Many were especially concerned by the fact that CG&E, and not the Nuclear Regulatory Commission, would be the one monitoring Zimmer's discharges into the Ohio.

The rest we know: from city council's direction, an assistant city solicitor (William Heile, you'll recall) filed a petition in 1975 with the NRC for the city to be an intervenor in Zimmer's upcoming licensing hearings and the following year secured that right. With federal and state reviews of the plant ongoing, it would be another three years, not until the spring of 1979, before licensing hearings commenced and intervenors got their chance to fully participate in that government review.

Most people recall Jerry Springer for his profanity- and scandal-filled tabloid talk show in the 1990s, but that eclipses—and was, frankly, a complete departure from—the important political career he had before that. In fact, his election to city council in 1971 was part of a power shift in Cincinnati that resulted in the city government becoming critical of Zimmer. Springer was a member of a new coalition of Democrats and Charterites (a progressive reform party), which displaced long-standing Republican leadership in city council. The coalition held a majority in council for the entirety of Zimmer, and since it was custom in Cincinnati for the mayor to be chosen by the majority party in council, mayors in these years were also Democrats and Charterites. (Cincinnati had—and has—a unique governance system in that there is a mayor, a city council, and a city manager.) In contrast to Republican predecessors and peers, Springer and other coalition members were generally younger, included women and Black Americans, cared about social justice issues including the environment and public health, and wanted to make local government more responsive to the public. As such, council Democrats and Charterites challenged CG&E over Zimmer's radioactive emissions—and they challenged the way the company operated: seemingly untouchable, without oversight.

And it wasn't just CG&E's actions around the licensure and operation of Zimmer that provoked Democrats and Charterites.

In the 1970s, CG&E frequently tried to raise electric and gas rates for city of Cincinnati customers, doing so through negotiations with city council. Springer and the coalition challenged these hikes, probing to see if corporate greed was at play. CG&E representatives insisted it was not, pointing to various issues to explain their financial predicament, from rising coal prices to new environmental legislation that required costly upgrades to power plants. They also made clear that two major developments—the 1973–1974 national shortage on oil and gasoline (the so-called energy crisis) and the 1973 recession, which ended the great post–World War II economic boom—were jeopardizing CG&E's ability to build power plants. All of that was true; these issues were dramatically impacting the utility industry across the United States. But the issue was that CG&E's projections of future electric and gas demand were a factor in determining their rate base amounts. Understanding this, Springer and other coalition members argued that the poor economy was curbing energy demand, so CG&E's plans for growth needed to be similarly curbed. In other words, CG&E wouldn't ultimately need those rate increases. As Springer and his colleagues wrestled with the utility's leadership, a local anti-CG&E movement emerged, where—with the bad economy—residents (especially working-class and low-income ones) protested rate hikes and CG&E customer service. Scrutinizing and criticizing the incurred costs that CG&E included in its rate base, they were part of a wave of anti-utility, consumer rights agitation that swept the US in the mid-1970s.

None of this organizing, in the minds of protesters, was explicitly about Zimmer. From CG&E's perspective, it was, though: as company executives expressed, the energy crisis and recession implicated their nuclear power plant because if they couldn't increase rates, Zimmer's financing and completion—as a large and expensive investment—were at stake. But few CG&E customers in the early and mid-1970s were openly attacking Zimmer, namely because its costs were not *yet* included in the rate base. A power plant had to be at least 75 percent complete for that to be the case, and even as CG&E embarked on Zimmer's construction immediately following its 1972 permit, it wasn't until the end of the decade that Zimmer was over three-fourths completed. Instead, the actions of CG&E customers and council Democrats and Charterites around 1973 laid a foundation—a foundation that would later, by around 1980, blossom into a powerful, local consumer rights dimension to the anti-Zimmer fight. In the meantime, as Springer and his coalition elevated Zimmer as a possible threat to public health, more people—than just those in Moscow, Ohio—began to follow news of the plant.

Jerry Springer was born in the Highgate underground train station in London during World War II when it was being used as a bomb shelter. His father, a shoe and toy wholesaler, and his mother, a bank clerk, were German Jewish refugees. Five years after his birth, they immigrated to Queens, New York, where they lived in a second-story walkup. Springer later went to Tulane University for his undergraduate degree, earned a law degree from Northwestern, and during law school clerked in Cincinnati at a law firm. Through this, he became—in his own words—"a liberal." "If you are a child of Holocaust survivors," he reflected later in life, "it's hard not to be a liberal. Twenty-seven members of my family were wiped out. You learn that you never judge people on what they are, but what they do." After law school, he served as Robert F. Kennedy's legislative aid when Kennedy was a New York senator, and he got involved in Kennedy's run for the Democratic presidential nomination before he was assassinated in 1968. Springer was always forthright about how much Kennedy's passion for social justice shaped his own.[2]

To this point, Springer ran for Congress in 1970 on an antiwar, pro-environment platform. Unsuccessful, he returned to Cincinnati and won a seat on city council in 1971. In his campaigning, we can see what kind of politician he was. He took a stand against "violence" in the city, which he clarified meant violence "that afflicts the poor, that poisons relationships between men because their skins are different colors, their parents' different backgrounds, or their churches' different prayers. It is the violence of a hungry child who might never laugh, or of an overcrowded school where he might never be taught." And so, once elected to council, he pursued wide-ranging reforms. He authorized a bill requiring more minority hiring in the police. He worked with the nonprofit Legal Aid Society to strengthen the rights of building tenants over landlords. He got the city to hire more building inspectors so that housing conditions, especially for poor residents, would improve. He also introduced an ordinance controlling the use of asbestos. And while he resigned in 1974 after admitting to soliciting a prostitute, voters didn't seem to care. They elected him back in 1975 as an independent Democrat.[3]

Jerry Springer—and his fellow Democrats and Charterites—disrupted over seventy years of Republican control of Cincinnati politics. Republicans took the lead in local government around 1900, when their party stood for city beautification and betterment while also ironically supporting a corrupt city boss. In the 1920s, a group of local Democrats and Republicans created a new party, the Charter Committee, to eliminate corruption and introduce merit-based leadership to city government. Charterites initiated at-large voting for a nine-person city council and the practice of city council choosing a "weak" mayor to preside—changes that persisted through Zimmer's timeline. But while Charterites were successful in passing a new city charter that embodied their goals, and while they continued to be a local force, standing for accountable government often in alliance with Democrats, Republicans gained control of city council again and held it through the 1960s.[4]

By then, local Republicans cared most about making Cincinnati "an area on the move," to borrow the phrase of their friends in CG&E leadership. In accordance, Republican officials pursued downtown redevelopment—like the convention center and Stouffer's Hotel—to try to stop the flight of residents, businesses, and tax revenue from the city. They struggled to understand the plight of African Americans and Appalachians, who were, by midcentury, the majority populations in the heart of Cincinnati. These groups, who moved to the city for better economic prospects than the South offered, arrived with few resources and encountered significant hardship and prejudice in employment and education. One way Republican leaders responded to growing inner-city poverty was through slum clearance—demolishing dense sections of housing—and the building of new segregated public housing. Tens of thousands of Black residents ended up being displaced. On top of this, when the local National Association for the Advancement of Colored People, the NAACP, fought to desegregate the city, Republican leaders called for *gradual* change and integration. And when racial unrest—riots, civil disobedience—broke out in the late 1960s, Republicans called for "law and order" rather than asking, "Why are Black residents upset?"[5]

Changing demographics in the city made these actions increasingly unacceptable. As white families fled to unannexed suburbs farther out in Hamilton County (and beyond), they tended to vote Republican there (leading to county government being in Republican hands). The incorporated city of Cincinnati conversely "turned blue." By the 1960s, Black Americans constituted a major voting bloc, and they joined with progressive white voters to elect Democrats and their allies, the Charterites; these parties were showing themselves to be supporters of causes that mattered to these constituents—causes like racial justice, environmental protection, and women's rights. This was occurring in other midwestern and East Coast cities, too, where a new generation of leaders was elected because they cared about these social issues.

At the national level, the progressive wing of the Democratic Party spearheaded this, as the presidencies of John F. Kennedy and Lyndon B. Johnson showed. Under Johnson, federal funds flowed to city leaders for antipoverty programs and other initiatives to make cities more equitable and livable. The aid, which mandated citizen participation in planning and implementation, fostered a resurgence of "people power" at the local level, as was obvious at Zimmer's public hearings. Residents in Cincinnati and other cities were tired of the top-down city leadership of the 1940s and 1950s, where decisions were made about neighborhoods over the voices of people who lived there. Voters wanted a voice in local decision-making, and in Cincinnati, Springer and his coalition were sensitive and sympathetic to this.[6]

In 1969, the Charterite-Democratic coalition secured four out of the nine seats on city council. In 1971, it took a majority and held that until 1985. A striking example of the coalition's popularity and people's civic engagement was that almost 70 percent of registered voters in the city of Cincinnati went to the polls in the mid-1970s, to date the highest turnout.[7]

With the change in leadership, city council looked different—literally. It had its first Black mayor in the 1970s, Theodore "Ted" Berry; its first full-term female mayor, Bobbie Sterne; and some of its youngest councilors. Springer was only in his late twenties on election to city council. James Cissell, another Democrat, who had served as the assistant attorney general for Ohio prior to his election to council in 1974, was only three years older than Springer. All of these people would challenge the Zimmer project in important ways.

Ted Berry was an indefatigable civil rights activist who came to the mayor's office with significant experience in antidiscrimination, antipoverty, and government work. He was born in Maysville, Kentucky, in 1905 and moved with his family to Cincinnati's West End as a young child. He grew up in stark poverty. His mother (who was deaf and mute) eked out a living doing other people's laundry. Nonetheless, showing his perseverance and drive, Berry was the first Black valedictorian at Cincinnati's Woodward High School. He graduated from the University of Cincinnati in 1928 with his bachelor's, after paying his way through multiple manual labor jobs, and then attended law school there as the only Black student in his class. After being admitted to the Ohio Supreme Court Bar in 1932 and the US Supreme Court in 1937, he held different, groundbreaking appointments, such as being the first Black assistant prosecutor of Hamilton County. His legal work became intertwined with racial justice. He led the Cincinnati chapter of the NAACP from 1932 to 1946 and then served on its national board until 1965, in which capacity he sued Crosley Radio, a major Cincinnati manufacturing company, for not hiring Black workers. He also took on the local Board of Education for having no Black members. And he provided legal counsel for national civil rights cases, including defending the Tuskegee Airmen in 1945. (Three Black Army Air Force officers who had served in World War II were being court-martialed for violating racial norms—for allegedly shoving a white officer and, along with 101 other Black officers who had also been arrested, for refusing to accept that they weren't allowed in a white officers' club in Indiana. Berry secured the release of all men from jail and got acquittals for two of the men accused of shoving.) He then moved into government work: in 1965, President Johnson appointed him to manage federal programs on civil rights and antipoverty work. Four years later, following Nixon's election, Berry returned to Cincinnati, where he immersed himself in local politics. Having served on city council in the 1950s and early 1960s as one of its first Black members, Berry was a familiar success story, and the people elected him to city council in 1971. From there, he served as mayor from 1972 to 1975.[8]

There was also Bobbie Sterne, Berry's Charterite colleague. She was one of the first female mayors of a major US city. Born Lavergne Mary Lynn in 1919 in a rural northeast Ohio town, she grew up poor during the Great Depression, although she was known to be a "glass-half-full person." Her male nickname of Bobbie came from childhood when her parents taught her she could do anything a boy could. And her life personified that. She studied to become a registered nurse and then joined the Army Nurse Corps during World War II, serving on battlefields in Europe. Afterward, she and her husband (an army doctor) settled in Cincinnati, where—as a wife and mother—she became active in civic matters. First, it was Girl Scouts and PTA, courtesy of her daughters. Then she became active in the League of Women Voters, the Woman's City Club, and the Charter Committee, all organizations that cared about women's rights and accountable leadership. With her medical training, she also volunteered for the city's Department of Health and Human Services, doing door-to-door work in Cincinnati's poorest neighborhoods to make sure children received vaccines. All of this predated her city council terms, which were from 1971 to 1985 and again from 1987 to 1998 (and in 1976 and again in 1979, she was chosen as mayor). In office, she was the same Bobbie: well-dressed, reserved in demeanor, but emboldened to do what she thought was right. She always prioritized public health, human services, education, women's rights, and the environment. Among other campaigns, she improved the city's affirmative action hiring, worked to end sex-segregated want ads in local newspapers, and declared Cincinnati's first gay pride day.[9]

Sterne's Charterite-Democratic coalition confirmed new city managers (who oversee all city administration) with a similar fever for change. Springer said as much in 1973: "Coalition leadership brought a new outlook to government in Cincinnati. We sought and secured a new city manager who was sensitive to people and had both professional training and professional experience in dealing with neighborhood problems." City manager from 1971 to 1975, E. Robert "Bob" Turner—then the youngest city manager—had "little respect for desk-bound bureaucracy" and "the tired excuse of 'that's-the-way-it's-always-been,'" as *Cincinnati Magazine* wrote in 1972. "I think I have been a consistent advocate of change and will continue to be," Turner stated at the time. He believed that people "who stay too long [in city government] get locked in by traditional attitudes." The next city manager, William Donaldson, also stood for accountable and accessible government. To open city hall to citizens, he literally removed the doors to his office.[10]

With its diverse and young energy, the coalition did everything it could to make local government accountable and its decisions transparent. It established mobile city halls throughout different neighborhoods to better reach citizens. It held meetings at night, enabling more people to attend. It brought community representatives into the city's budgeting process. One initiative in

1973–1974 required each city department to meet with neighborhood community council representatives to review departmental budgets. Another of its actions was to establish the city's first environmental task force—which would have something to say about Zimmer.[11]

In the early 1970s, Cincinnatians could expect to read something daily about the environment in their local news (which a mere decade earlier would have been unheard of). The *Cincinnati Enquirer* had hired Jo-Ann Albers and then Ben Kaufman for its new "environmental beat." The *Cincinnati Post* had Richard Gibeau as its environmental reporter along with Ron Liebau. The increased coverage underscored people's growing interest in environmental issues, and for Cincinnati, the coalition on council wanted to confront those issues head on.[12]

Created in May 1972, the city's environmental task force was both a discussion forum and an advisory board for the city. It was organized into topical subcommittees, which each held public hearings, prepared reports, and, by May 30, 1973, made final recommendations to council. Twenty-seven volunteer citizens, all appointed by council, were tasked with researching topics to help the city better provide—in the words of the city ordinance—"clean air, pure water, the scenic, natural and aesthetic qualities of his environment, and freedom from excessive noise." City council appointed women and men from a variety of professional backgrounds, so there were lawyers, scientists, corporate executives, engineers, professors, urban planners, and physicians involved. Some people were already active in environmental protection around the city, and of course by volunteering for such a board, task force members indicated that they clearly cared about the environment—whatever that meant to each person. It wasn't surprising, then, that many of the task force's recommendations were urgent and bold, and many specified a heightened role for state and local government in monitoring the environment. That said, the group saw people debate the specifics, like how far to take recommendations and how much the government should be involved in enacting changes.[13]

This was most apparent in the energy subcommittee. Selig "Ted" Isaacs chaired the group. The son of Jewish immigrants from Lithuania, he was a chemical engineer, the president of an industrial equipment supplier, a lover of the arts, and by the time of the task force, almost sixty years of age. Lawrence Kessler—the director of the environmental group, the Tri-State Air Committee—also sat on the energy committee. (He was the one who declared the Atomic Safety and Licensing Board hearings that year a "fraud" and a "sham.") There was also a radiologist along with two women involved in local natural conservation and other civic matters. The group's public hearings, held in city hall council chambers, attracted city residents as well as people from Clermont County, where Zimmer sat. (David Fankhauser came.) Leaders from local environmental groups also participated. CG&E representatives agreed to attend.

As much as the subcommittee members endeavored to keep things orderly, the sessions often devolved into contentious meetings about nuclear power, to the exclusion of other energy-related topics. Like with the Atomic Safety and Licensing Board hearings, where Fankhauser dueled with other scientific experts over the safety of nuclear power, the task force hearings heard from technically trained people on both sides of the fence. At one meeting in late 1972, nuclear energy critic Richard E. Webb told the audience that there was no such thing as "a fail-proof system," where multiple barriers around a nuclear reactor would prevent a major release of radiation. Webb was a nuclear engineer who helped to design the nation's first commercial nuclear reactor at Shippingport, Pennsylvania. In response, University of Cincinnati nuclear engineer James H. Leonard disagreed, going as far as saying that Zimmer's steel-lined concrete dome "will withstand a Boeing 707 crashing into it or a hurricane."[14]

The tension was palpable, even to the press. "More emotionalism surfaced at [the energy hearing] than at the other five [other task force] hearings combined," the *Cincinnati Enquirer* reported afterward. (For context, other subcommittees deliberated on topics including air pollution, landfills, asbestos, and historic building designation.) Isaacs's energy subcommittee likewise noted how "strongly divergent opinions on safety and environmental effects of nuclear power were presented. Since many of the people who appeared at these hearings have what seem to be equal expertise . . . the only possible conclusion is that the safety and environmental effects are not fully understood at this time."[15]

In the midst of the subcommittee's work, Ernest Sternglass published his controversial paper on the Shippingport reactor, where he attributed a spike in cancer rates in the Ohio River Valley to it. Sternglass was a German Jewish refugee whose family fled to the US in 1938. He studied engineering at Cornell, served in the navy, and then worked in government-contracted labs, engineering new X-ray and light technology to the tune of thirteen patents on radiation devices. While at the Westinghouse Research Laboratory, he cocreated the highly light-sensitive camera and photomultiplier tube, which NASA adopted for several space missions, including the Apollo 11 moon landing in 1969. (So he's responsible for the camera that captured that moment.) In 1967, he began work as a professor of radiation physics at the University of Pittsburgh, where he led the team that developed the first digital X-ray systems. By then, he was staunchly opposed to nuclear weapons and nuclear power plants. He even testified before the Joint Committee on Atomic Energy in 1963 regarding the strontium-90 in children's teeth as a result of atmospheric nuclear bomb testing (which was banned that year). Like Fankhauser, he believed that "every dose of radiation carries a finite risk." He didn't trust the permissible emission levels for nuclear power plants set by the Atomic Energy Commission and the Nuclear Regulatory Commission, so he compiled large datasets on emissions and disease/death rates near reactors. His Shippingport paper was among many of his that linked heightened mortality outcomes to casual radiation exposure.[16]

Sternglass wrote that Shippingport's reactor was contributing to "excessive amounts of radioactive fission products in the soil, the milk and the air of Western Pennsylvania," which he argued affected downstream and downwind populations. Using samples of surface water from 1964 to 1972 supplied by the Pennsylvania Water Quality Network, he compared concentrations of radioactivity from various Ohio River locations upstream and downstream of Shippingport. Then, after looking at cancer death rates from 1958 to 1968, he concluded that radiation levels around the plant "exceeded by some 50,000 times the levels officially reported by Duquesne [the utility in Pennsylvania] to the Environmental Protection Agency, including high levels of strontium-90 in soil and milk that corresponded to rises in monthly electric output of the plant." He reported that an "abnormally high rise in cancer mortality reached all the way to Cincinnati." When he visited the University of Cincinnati's student center in the spring of 1973, he told students and professors there, "You are drinking the effluent from Shippingport."[17]

Because his data affected the city of Cincinnati, Sternglass wrote to mayor Ted Berry and urged him to fight the Zimmer plant. Sternglass said, as a way of encouragement, that "the city of Pittsburgh has joined a number of environmental organizations in an intervention against two large new reactors proposed for the Shippingport site on the Ohio." Concerned, Berry met with him and afterward sought additional opinions on the report.[18]

Reinforcing his concern, Berry's office received a volume of letters from worried residents. One woman, a mother, told Berry that she and others were worried about the "water investigation." "This latest alarming report," she wrote, "coupled with all the other recent reports of cancer-causing sodium nitrite (used as a preservative) in so very many of our common foods; particles of insects and other dirt in many of our canned and packaged goods, causes me to wonder what in the world I can do to avoid poisoning my family or exposing them to dangerous elements through our food and drink every single day—in countless ways! Who, if not you and our governmental protective agencies, can protect us from all these dangers?" While her letter was a little overblown (insects in canned goods?), it does show that people were paying attention to "dangerous elements" in their environments. And the fact that she was looking to her mayor for protection shows us that Berry and his coalition were trying to be more accountable, more responsive—and people were noticing.[19]

In response to the unfolding issue, the Ohio EPA, Ohio Department of Health, and the Ohio River Valley Water Sanitation Commission studied Sternglass's data. Ultimately, though, they rejected his conclusions. Ohio EPA and the Health Department concluded that the incidences of cancer that Sternglass cited did not account for the age, sex, and size of the population being studied. They also found "no evidence of any increase in genetics defects or infant mortality." The Atomic Energy Commission issued its own rebuttal to Sternglass, discarding his findings as cherry-picked: "The water quality data on gross-beta radioactivity has merit as a long-term indicator of trends, but individual periodic sample data cannot be used to make quantitative estimates of radioactivity releases from specific sources, as was done by Dr. Sternglass." To assure the public of Zimmer's safety during the Sternglass controversy, CG&E wrote an op-ed in the *Cincinnati Enquirer*: "We are making certain that the plant will operate below the allowable limits of emissions, and will be a safe and clean asset to the community." Summing up the controversy to mayor Berry, city manager Bob Turner wrote, "Dr. Sternglass selected values from statistical tables to prove his contention. From the same tables you may select numbers of values to disprove it." He added, though, "Not enough is known about the long-range effects of even very small amounts of radioactivity."[20]

At the time, Sternglass offered his opinion on why so many scientists attacked his report. "It's hard to find someone in the scientific community who hasn't worked for the Atomic Energy Commission," he told Jo-Ann Albers at the *Enquirer*. It was true that the AEC and its affiliated scientists had monopolized research on radiation until then—although by the 1960s and the 1970s, there were some nuclear physicists who were beginning to challenge the AEC and later its successor agency. Some of them even quit their government lab jobs, formed antinuclear organizations, and published books and papers on the topic. One Nuclear Regulatory Commission official named Robert Pollard, who reviewed nuclear plants for licensure, very publicly resigned in 1976. He was convinced that the NRC lacked the regulatory ability to protect the public from danger. Pollard then joined the Union of Concerned Scientists, a national organization founded in 1969 by scientists and students at the Massachusetts Institute of Technology who were appalled by how, in their opinion, the US government was misusing science. Members became watchdogs of the commercial nuclear power industry. All this is to say that Sternglass was far from alone.[21]

To this point, his report—even with the issues raised about it—still forced Cincinnati to pause. The city's Water Works issued a statement that radioactivity in Cincinnati water had not reached a dangerous point, but it still found it prudent to note that "this report is not intended to say that everything is under control and we need not worry." University of Cincinnati radiologist Eugene L. Saenger originally snubbed Sternglass's findings as "scientifically incorrect" since data from the National Cancer Institute and National Institute of Health did not show uniformly increased rates of cancer mortality in communities downstream of Shippingport. But he later backtracked and asked mayor Berry to urge the Ohio state assembly to create an agency to monitor radioactivity. After all, Saenger said, the state had a number of nuclear power plants proposed or already underway (by 1976, four, in fact).[22]

The Sternglass paper impacted the city's environmental task force. The energy subcommittee's final report to city council said that while fossil-fueled plants "generate well known and relatively serious air and water pollution," and while nuclear power "does not produce pollution that people can sense," "it does discharge radiation into the air and water. There may be both long term genetic effects and possible accidents that could destroy all life within a 100 mile radius of the plant." Subcommittee members also noted that a massive accident through human error was possible, writing that "some scientists insist there is real and compelling danger of accident such as loss of coolant or through human error." Subcommittee chairman Ted Issacs commented that his twenty-five years of selling industrial equipment—including some of the very control valves used for Zimmer's reactor—had convinced him that a mechanical failure in nuclear power plant control systems was a "virtual certainty."[23]

With nuclear power's unknown health ramifications, the subcommittee concluded that "the City of Cincinnati should exert every effort to forestall operation of the Zimmer Power Station." As a part of that, it stressed that the city should become an intervenor in Zimmer's licensing hearings to ensure the plant was built as safely as possible. Unconvinced that CG&E would do a good job of monitoring emissions from Zimmer, the subcommittee told the city to establish its own radiation monitoring department. It also concluded that the Price-Anderson Act, which limited utility liability in the event of a nuclear power plant accident, was unfair: "CG&E will not bear the financial liabilities in the event injury to citizens and damage to their property occurs."[24]

City council also received a "minority report" from other task force members, a group of engineers and nuclear physicists from local universities, who, after reading the energy subcommittee's assessment, took issue with it. They said it was "unduly pessimistic" and displayed "what appears to be a strong bias in their deliberations that existed from the onset of their activities." By this, they meant a strong *antinuclear* bias and reminded city officials of the "careful research and planning that have gone into the development of nuclear reactors and nuclear power plants." To them, "overwhelming expert testimony regarding the safety of such plants" should not be ignored.[25]

City council noted this disagreement, which resembled the same tension between scientists that had occurred at the Atomic Safety and Licensing Board hearings. Furthermore, city manager

Bob Turner wanted the city to get involved with Zimmer, but instead of intervention, he recommended a more neutral path: city council could create and send position papers to the Atomic Energy Commission, which would, ideally, be used in licensing hearings to advocate for the city's public health. With these various recommendations in front of them, Charterites and Democrats on city council asked their new Environmental Advisory Council for advice. In 1972, as the environmental task force deliberated, the city had created the advisory council to take the place of the environmental task force after it disbanded. Its members were selected by the city manager for two-year appointments. Responding to the Zimmer situation, the Environmental Advisory Council read through the task force's reports on Zimmer and CG&E's application for a construction permit and ultimately arrived at the same point that Jerry Springer had expressed in 1972—that Zimmer would discharge radiation into the Ohio River and that CG&E alone would monitor this. The council recommended the city intervene in Zimmer's licensing hearings, and the Charterite-Democratic coalition followed through on that.[26]

The few Republicans on city council, who were in a minority position, disagreed. They argued that intervention was excessive involvement by the city; that CG&E, General Electric, and the Atomic Energy Commission knew how to build and regulate Zimmer, including keeping the city's water safe; and that Zimmer was urgently needed to meet the region's electricity demand. In this, local Republicans were saying the same thing as CG&E executives, revealing their close relationship. Perhaps most telling, one of the Republicans saying these things was former CG&E executive Walter E. Beckjord. The son of Walter C. Beckjord (CG&E president from 1945 to 1957), Beckjord, the junior, served as the company's vice president and general counsel before his city council service.[27]

In 1976, Zimmer again revealed these partisan lines. As the Nuclear Regulatory Commission was granting the city's request for intervention that year, Springer introduced a motion to ban air and ground shipments of radioactive material going through the city, following a similar move by New York City's health department. Zimmer's uranium fuel was to come by truck from Wilmington, North Carolina, and the question was whether it should travel through the city's corporation boundaries en route. Springer told his fellow councilors, "Transportation of nuclear materials through our streets creates the potential for catastrophe." His colleague James Cissell added that an accident "could pose serious consequences for generations yet unborn, and could cause the earth to be uninhabitable." Springer's proposed ordinance further required any company transporting nonradioactive but otherwise dangerous substances through the city to have city-issued certificates of transport. While the US Department of Transportation had regulations covering many of the materials, Springer insisted, "The public has to make the decisions about what kinds of materials they want in their communities."[28]

The motion encountered stiff opposition from Republican city council members and Republican officials outside the city of Cincinnati. In New Richmond, close to Zimmer, one village council member there called Springer's motion—and, in fact, all city of Cincinnati involvement in Zimmer—a "witch hunt" designed to hurt CG&E's ability to build and operate Zimmer. In response to the claim that radioactive discharges from the plant would contaminate Cincinnati drinking water, the man charged, "I think it's the other way around. Cincinnati is polluting us with their smoke and pollution. We're not going to pollute them."[29]

The conflict around Springer's motion worsened when he tried to convene a public hearing on the matter, hoping a panel of scientists could advise on the issue and, importantly, answer people's questions. He invited three nuclear engineers who had recently quit General Electric after finding safety deficiencies within a number of nuclear power stations. On resigning, they expressed that the commercial nuclear power industry was rife with such issues. Republican council members balked at these speakers, questioning their credentials. Walter E. Beckjord asked, "I've read they are qualified engineers for nuclear plants. Has anybody said they are experts in transportation of nuclear materials?" Republicans succeeded in postponing the hearing, and soon after, the motion died in committee, considered (even by those who believed in it) too difficult to implement.[30]

Despite the conflict in city council, partisan lines around Zimmer were not as clear-cut as they might seem. Part of that was because partisan lines around nuclear power weren't clear-cut. Look at the presidents in these years: Nixon, a Republican, supported nuclear power and called for significant investment in it. So did Gerald Ford, his Republican successor, and so did Jimmy Carter, the Democratic president who served from 1977 to 1981. Carter embraced the promise of nuclear power in part because of a major oil shortage and other energy issues during his term. But he also backed nuclear power because he was familiar with the technology (he was a trained nuclear engineer who had worked on a nuclear submarine under Admiral Hyman Rickover, the "father" of the navy's nuclear program) and because Carter considered himself an environmentalist. As such, in the 1970s, there were some progressive Democrats who supported nuclear energy, seeing it as a cleaner alternative to fossil fuels like coal. At the same time, there were progressive Democrats who opposed it for its public health and ecological risks. On the other political aisle, there were Republicans who opposed nuclear power. Holding conservative family values, they objected to the reproductive health risks associated with nuclear power station operations. Property-rights conservatives didn't always like nuclear power, either, because a nuclear power plant could destroy the surrounding area's natural beauty, reducing home values. So, for Cincinnati, it's incorrect to say that all local Republicans supported Zimmer, just like it's erroneous to say all Democrats and Charterites opposed the plant.[31]

Relatedly, despite the "witch hunt" claims against the city by the New Richmond official, the battle over Zimmer could not be reduced to the city of Cincinnati versus the suburbs and rural areas around the city, where the latter wanted the power plant and the former didn't. Fankhauser's activism underscores the opposition from people close to Zimmer. Springer's work from city hall—miles away from Zimmer, outside of city jurisdiction—points to the opposition from people far away from the power plant. And moving forward in time, the opposition would only diversify, with more people from the rural, conservative area around Zimmer opposing the project. But that's jumping ahead.

A final word on Springer's motion to regulate the transport of radioactive substances: it died in committee but lived on elsewhere. In fact, by 1979, eleven states and eighteen local governments had successfully placed restrictions on the intracity transport of nuclear material. By 1980, that number had risen to one hundred, with states, cities, and counties drawing language and substance from New York City's regulation. After the Department of Transportation tried to overrule these local laws in 1980 by designating interstate highways as the main routes for shipment of radioactive waste, New York City sued the federal agency. Ultimately, after several appeals, the Supreme Court upheld the DOT's regulation.[32]

Still, the nationwide action by elected officials to reroute nuclear material away from their populations underscored a new attitude among many politicians and officials in the 1970s that had implications for Zimmer and other proposed nuclear power plants. With public health and the environment in mind, many leaders erred on the side of caution, and they sought additional accountability from governmental agencies to ensure public safety. This was why New York state governor Mario Cuomo challenged and then decided to fight Long Island's Shoreham nuclear power plant in the 1980s and why California's governor from 1975 to 1983, Jerry Brown, was critical of nuclear power stations. He and the state's energy commission, created by voters in 1974 to oversee state energy development, fought many proposed nuclear power plants, questioning utilities' need for them and their public health implications.[33]

In Cincinnati, as Springer and his coalition questioned Zimmer's impacts on citywide water, and as the *Cincinnati Enquirer* and *Post*'s reporters covered the power plant, people—and not just those in Moscow, Ohio—were forced to consider its potential safety impacts. Aside from this, something else percolated in the mid-1970s that *really* got local people to have an opinion on CG&E and its arsenal of construction projects: rising utility rates.

In Ohio, electricity and gas pricing was and is set by the Public Utilities Commission of Ohio (usually shortened to the PUCO), which is staffed by five commissioners appointed by the governor. Customers' monthly charges are determined by rate prices, or bases. For electricity, utility users can see this rate base listed on their bills as "so many cents per kilowatt-hour of energy" and for natural gas as "so many cents per therm" (a therm being a measurement of the amount of heat energy in natural gas).

Rate pricing includes and reflects a variety of costs that utility companies carry to produce power. In the 1970s, the PUCO allowed utilities to recover part of their electric and gas fuel costs by including those costs in customers' rate bases. Since fuel costs frequently vacillated, the PUCO allowed utility companies to automatically pass minor increases or decreases in fuel costs to their customers. These were known as fuel adjustments. For major increases in rates, a utility had to obtain approval from the PUCO first.

Like all state utility commissions, the PUCO supported the idea that a power company had a right to earn a reasonable rate. Beginning in the early 1900s, utilities could charge customers rates that would allow them to collect a fair return on all that they had put into producing power. Rates, then, were a factor of utilities' total investment needed to create power (all their past and present incurred costs) and how much energy they expected people to use, using consumption data from past years. (An important caveat is that in Ohio and certain other states, a power plant under construction had to be at least 75 percent complete for its incurred costs to be included in people's rates. Furthermore, when a utility included that work, it could not raise customers' rates by more than 20 percent of their original rate.) So, to customers, a rate base looked like a simple number on their bills, but in reality, that number encompassed their power provider's fuel costs, other incurred costs related to power production, and expected power production costs for the future.

When utility companies desired higher rates to reflect higher costs (expected or real), they presented proposed increases to state utility commissions. These commissions, like Ohio's PUCO, approved, changed, or denied them, although utilities generally won the right to put into their rate bases most of their incurred costs. (The verbiage that the PUCO used in granting a rate increase was often "CG&E is authorized to an increase in annual electric revenue to the tune of so-many-millions of dollars," which was another way of saying that CG&E had received an electric rate increase that would result in additional annual revenue for the company.)

State utility commissions were supposed to be staffed with politically neutral experts, but poor salaries and limited funding meant regulators didn't have the resources to contest claims of utilities. (For reference, in 1974, the chairman of New York's Public Service commission earned $51,150; in comparison, the president of New York City's utility, Consolidated Edison Company, made $211,000.) Because power companies didn't want the commissions to appear illegitimate, utility executives often hyped up the credibility of state utility agencies, creating the impression that the regulators protected consumers, while in reality they had a cozy relationship with utilities. Furthermore, in 1944, the Supreme Court ruled that rates power companies charged didn't just have to be high enough for them to earn a fair return—but high enough to maintain a good credit rating and be able to compensate shareholders. In other words, utility customers had to pay rates not just high enough to make sure utility providers made money but that utilities' investors did as well.[34]

For CG&E customers *within* the city of Cincinnati, who made up one-third of CG&E's ratepaying base, the process for rate increases worked a little differently. CG&E still had to justify its costs and its consumption predictions for future years, but instead of presenting this information to the Public Utilities Commission, CG&E first negotiated with Cincinnati's city council, and on agreement, new rates for municipal customers were made official in a city ordinance. (CG&E did have the right to ask the PUCO to intervene if negotiations stalled with city officials.)[35]

Once any rate changes were approved, for the city of Cincinnati or otherwise, the Federal Power Commission had to approve them. Since the review process for any and all rate changes took a while, CG&E and other utilities complained that rate increases were slow to take effect, and once they did, that they were outdated—based on costs incurred for the year or two prior. For instance, CG&E requested an electric rate increase in 1970 and received it in 1973. Still, the rate review system benefited utilities. It was sympathetic to the enormity of costs that went into power production, and it protected power companies from competitive market forces because rates weren't calculated based on what the price of electricity "might fetch." Up to the 1970s, customers accepted this system because they experienced low electric and gas rates. And even though utilities' rate bases were based on costs incurred and even though utilities engaged in a construction spree after World War II, new technology and other improvements kept making electricity generation more efficient, lowering utilities' unit costs when they installed more power plants.[36]

This system of cheap rates and good returns began to fall apart in the early 1970s. Even before the 1973 energy crisis and recession, CG&E executives complained that various issues were hurting their company's finances, resulting in (in their opinion) much-needed rate hikes. One bigger-picture problem was that as power companies encouraged manufacturers to produce increasingly large turbine generators, the efficiency and reliability of the new generating units began to plateau, and only at tremendous construction and operating costs could they be made better. Since the PUCO required that CG&E supply continuous power—a condition given to all utilities by state commissions—CG&E was incentivized to rely on its less efficient but more reliable plants.[37]

CG&E officials also lamented other issues affecting their operations and finances: the sale price of natural gas to customers was too low, and on top of that, there was a gas supply shortage; coal prices were rising; and new air and water pollution regulations—which affected both coal-fueled plants and Zimmer—were causing costly upgrades. CG&E officers weren't exaggerating. Related to the last point, a new mandate, over how much the Zimmer plant needed to cool water before returning it to the Ohio River, added an unanticipated $11 million to the project budget. Company executives wrote city council in 1972, "We hope that our customers will understand that environmental enhancement, like everything else, has a price tag attached." With these various issues, CG&E terminated its rate ordinance with the city of Cincinnati in 1972, only two years after settling it, and started negotiations for higher electric and gas rates.[38]

Ted Berry and his coalition pushed back. It wasn't that they didn't believe the energy issues impacting CG&E, but rather they were fastidious in examining what CG&E sincerely needed. They wanted the company to make a fair, but not ridiculous, return. In talks with the utility, city councilors learned that CG&E passed on certain administrative and advertising costs to customers in their rate base. To CG&E executives, these were fair practices: they were allowed to include all incurred costs related to the production of power in the rate base, and wasn't administration required to make sure this happened? Furthermore, since they anticipated future electric demand using past consumption data, and since they had long had a hand in creating that demand through advertising, advertising costs—in the opinion of company leaders—belonged in the rate base. Mayor Berry disagreed, especially considering how people were sending him letters saying that CG&E's rate increases were "unnecessary luxuries" attributable to "advertising, high salaries, costly building, etc.," as one couple put it in 1972. Council members asked CG&E if it could instead enact energy conservation measures to save money. CG&E representatives said no. That, in no way, fit within their model for producing power. Berry pressed more, insisting that any new rate agreement between CG&E and the city had to include a requirement that CG&E improve its minority hiring.[39]

Republican council members recoiled. They asserted that a rate increase negotiation was not the time or place to talk about affirmative action, and moreover, CG&E needed and deserved the rate relief. After all—Republicans said—if CG&E's finances suffered, it wouldn't be able to provide sufficient and reliable energy for Cincinnati. In fact, it might not be able to finish Zimmer. Local business leaders were concerned about this too. A representative of the Greater Cincinnati Chamber of Commerce told city council, "If the new atomic plant is not 'on line' by 1977, other more expensive and more pollutant types of energy conversion will have to be provided." He continued, stressing Zimmer's importance, "We forecasted ability to meet area energy demand through 1977 with present and announced capability (including the Zimmer facility)." In 1972, the chamber's projections for regional economic growth showed a sharp upward curve and underscored that electric demand was expected to exponentially rise each year from 1972 to 1980, making Zimmer critical for meeting area energy needs.[40]

But Ted Berry, Jerry Springer, and the others in the coalition had begun to doubt those projections. Unconvinced that electric and gas demand would continue to precipitously rise, they demanded from CG&E their "long-range plans" and projections for the "immediate future." Furthermore, Berry reminded his council that the public had to be informed about what was going on: "The public has no information as to the allocation of the available surplus of power to meet the needs of continued growing demand. The prospect of power shortage in the near future is currently rumored. We need to know the facts to allay such rumors." Finally, after negotiating down as much as possible—from granting CG&E $11.3 million in additional revenue to $7.1

million—Democratic and Charterite council members approved new city rates with CG&E later in 1972.[41]

Only two years later, CG&E terminated its contract with the city to renegotiate higher prices again. This time, in addition to the preexisting issues stressing CG&E, there was also a national "energy crisis," which triggered a serious economic recession.

In October 1973, following American support for Israel in the 1973 Arab-Israeli War, Arab members of the Organization of Petroleum Exporting Countries, or OPEC, placed an embargo on oil for the US. Since the US burned oil for almost half of its energy consumption, particularly for transportation, and since it imported over one-third of its oil, the embargo dramatically curbed supply at gas stations and affected people's ability to power homes and businesses. In the 1960s and early 1970s, prior to 1973, utilities had begun to use more oil for electrical output since it was cheap and contained fewer impurities than coal. By 1973, almost 20 percent of electricity came from burning petroleum. So the energy shortage quickly manifested as a crisis.[42]

In Cincinnati, where CG&E only relied on oil during peak loads, the shortage was most felt as people struggled to fill up their gas tanks. As a result, people went to their garages and dusted off their bicycles, cycling to work, even through winter. Mayor Berry and his administration developed a program for conserving gas in city operations, including limiting the use of municipal automobiles, lowering city buildings' thermostats, and turning off unused equipment. City council contemplated lowering speed limits, and then Congress did this nationally in 1974, capping them at fifty-five miles per hour. Local companies tried to do their part. For example, Shillito's Department Store—Cincinnati's oldest—dimmed light levels in its downtown store and curtailed its use of Christmas lights. The embargo ended in March 1974, but oil prices remained high afterward, which, for CG&E and other coal-dependent utilities, caused coal prices to rise as well.[43]

Cheap oil had been king of the postwar boom of the 1950s and 1960s, so its shortage and subsequent high price pushed the

US economy into a recession in 1973—its first in decades. It effectively ended the postwar economic boom of 1947–1973, which had been the most sustained period of economic growth in world history. In 1973 and the next year and a half, Americans experienced declining real wages, growing interest rates, creeping inflation, and increasing unemployment. The stock market lost nearly half its value. For reference, in 1964, inflation was around 1 percent and unemployment 5 percent. In 1974, inflation was over 12 percent and unemployment above 7 percent.[44]

The poor economy called into question the entire business model of the utility industry. It was premised on anticipating and creating more demand, but as the economy went downhill, people conserved their money, curbing their electric and gas use. This contributed to lower sales for utilities. The utility industry was also structured around the idea of "build big, and efficiency will come of it." But inflation drove up fuel, materials, and labor costs, making large power plants even more expensive.

And what about nuclear power, the most expensive of power plants to build? The recession called it into question, too, especially when there were already growing delays in the nuclear power regulatory process. That it took seven or more years to move a plant through the licensure process meant that, in the meantime, during a recession, utilities like CG&E were burdened with holding costs (insurance, taxes—money you're just paying out without getting any kind of return). As such, power companies began to shy away from nuclear energy. Between 1966 and 1968, utilities purchased sixty-eight nuclear reactor units. Between 1975 and 1978, only eleven were ordered.[45]

But CG&E was one of the companies that had *already* ordered reactor units and begun construction, and in the midst of the energy and economic crisis, company officials did not suggest canceling or converting Zimmer to another fuel. Why would they? CG&E had already sunk significant money into the plant, which disincentivized it from backtracking (for reference, construction was already costing Zimmer's owners $640 million, about $240 million over their original budget). On top of that, company leaders still believed that Zimmer was needed. CG&E president William H. Dickhoner told his shareholders as much in 1975: "Our society's need for energy will increase in the future. The utility industry is capable of meeting the future demands for energy, but"—the one thing that would be different—"the years of cheap energy are past."[46]

At this point, in the mid-1970s, CG&E's rate base did not include Zimmer's costs. But the project was implicated in the turmoil of these years. After all, to finance the rest of Zimmer's construction, CG&E needed to be a healthy company, with high sales and revenue, so that its financiers—its stockholders, bondholders, and credit raters—were content and reassured. Higher rates, even if demand was curtailed, would help ensure higher sales. So that was CG&E's response to the energy crisis and the recession. In 1975, CG&E had applications pending before the Public Utilities Commission for rate increases to boost revenues by $81 million. In the two years prior, the PUCO had granted the company increases that caused the gas bills of residential ratepayers in southwest Ohio to rise by 50 percent. Customers also had their electric rates double in this short span.[47]

We can be sympathetic that higher rate prices helped CG&E combat the effects of inflation and provide an adequate return to investors. But leaders were stubborn to admit any responsibility for their financial predicament (like that they had had a hand in creating demand for electricity or that they had not compensated for how precarious the energy industry was). Their sidestepping, combined with the oil shortage and stagnant economy, made Americans very angry at their utility providers. Some even wondered if energy suppliers and power companies had invented the oil crisis just to drive up demand and prices. Local anger was apparent when two thousand people called in to CG&E *per day* in January 1974 alone, complaining about the rates. Clearly, there were plenty of Americans who merely wanted cheap and abundant electricity and gasoline to return. But there was also a growing

better monitor its use across Ohio. Meanwhile, after CG&E asked the PUCO for an electric rate hike for the city of Cincinnati, city solicitors—sent by council—traveled to Columbus, Ohio, and fought the issue at PUCO hearings. When they were unsuccessful, they appealed the decision to the Supreme Court of Ohio. When the state court agreed with the PUCO, the city appealed to the US Supreme Court.[53]

All of this anti-utility organizing in the early and mid-1970s laid important groundwork for Zimmer's fate later in the decade and into the 1980s. And the pattern that developed in those earlier years, where Democrats and Charterites went to bat against CG&E, persisted. When licensing hearings over Zimmer finally began in 1979, the coalition sent city solicitors to intervene, determined to ensure that Cincinnati's population would be safe from the plant. By then, this had become all the more urgent of a mission since in the spring of 1979, the US endured its first serious commercial nuclear power accident.

4

"Not Another Harrisburg"

Three Mile Island

Early in the morning on Wednesday, March 28, 1979, David Fankhauser left his farm to teach his classes at the University of Cincinnati's Clermont County campus. It was a spring day, cloudy but not too cold, as he headed north toward the college. Just a few miles in the opposite direction sat Zimmer. If he had driven past the plant on State Route 52, he would have seen Zimmer's cooling tower—a giant hourglass shape—shooting up into the sky. At that early hour, work crews were parked near the tower, outside the large building complex where the reactor and its containment vessel were. Inside, workers were installing hundreds of thousands of feet of giant electrical cables. At the same time, welders were structuring the vast piping systems, while another crew was readying itself to insulate those pipes with fire-rated material. By then, the spring of 1979, Zimmer's construction was already 91 percent complete, and CG&E expected it to be operational the following year. Fankhauser was still determined for it to be otherwise.[1]

Driving away from Zimmer, Fankhauser turned on the radio. Local newscasters informed listeners that one of the nuclear reactors at the Three Mile Island power plant in central Pennsylvania had had some kind of accident. How bad, no one knew yet. Almost forty years later, Fankhauser describes it as one of those days that sticks in your memory—a day, in his words, "when everything changed."[2]

Throughout that Wednesday, Metropolitan Edison—Met-Ed, the utility in charge of Three Mile Island—reported everything was fine. But as the week passed, it was clear that everything was far from fine. In fact, the reactor was experiencing a partial meltdown, and radiation was being released outside the plant. In the face of this serious situation, Met-Ed and Nuclear Regulatory Commission officials nonetheless insisted there was no public health threat. Later, when people had time to step back and assess, it would become known as the most serious accident that had ever happened at a commercial nuclear power plant in the US, a designation it still holds.[3]

"What does yesterday's nuclear accident in Pennsylvania mean to people living in Greater Cincinnati where—24 miles from downtown—the Zimmer nuclear station is nearing completion?" *Cincinnati Post* reporter Douglas Starr asked his readers. It meant that a growing number of people in the Cincinnati metropolitan area realized that what happened at the Three Mile Island station could happen at Zimmer—not necessarily the exact same sequence but a human error, a mechanical mishap. The unexpected, visible demonstration that nuclear power was not as fail-proof as promised caused residents across southwest Ohio and northern Kentucky to suddenly question Zimmer and seek a delay in its licensure—or even to demand its outright cancellation.

People petitioned local and state leaders to secure these outcomes. Unprecedented numbers also flocked to the plant's licensing hearings in downtown Cincinnati, which incidentally commenced only a few weeks after the accident at Three Mile Island. They were both Fankhauser types (people with activism experience or scientific training) and ordinary folks with working-class and middling incomes who didn't identify as activists. Mothers and fathers took center stage, as they had at the 1972 hearings. But a difference post–Three Mile Island was that residents were coming from *around* the metropolitan area, miles and miles away from Zimmer, rather than its immediate periphery. After all, they said, an accident at Zimmer wouldn't only affect Moscow. Radiation didn't stop at any imaginary border.

In the first outburst of "formal" organizing against Zimmer, the families who lived in the immediate zone around the nuclear power site created a group called Zimmer Area Citizens with the initial goal of ensuring more comprehensive accident preparedness at the plant. Other people joined Citizens against a Radioactive Environment, or CARE for short. A young activist named Tom Carpenter and a longtime peace and civil rights activist named Polly Brokaw—who was also David Fankhauser's mother—had started CARE just one year before Three Mile Island. They shaped it into an antinuclear power and pronuclear disarmament organization, reacting not only to the nuclear power stations being built across the US but also to the Cold War arms race that was then revving up again between the US and the Soviet Union. And while prior to Three Mile Island, their group could rally a few hundred people for events, an astonishing eight hundred people showed up to the first meeting CARE held after the accident. Then, in 1979, people were raising the same concerns over radiation that critics had in the early 1970s ("How likely is an accident?" "How dangerous is low-level radiation during operation?" "How safe are my children?"), but the major jump in CARE's membership shows the power of a crisis. It took Three Mile Island to make nuclear power's risks relevant to most Cincinnatians—and to most of America, which was evident in the outpouring of antinuclear marches in the days after.

In Cincinnati, the accident actually reversed some people's loyalty to Zimmer. Local Republican officials from around Moscow suddenly tempered their enthusiasm for the project. Worried about their constituents' health, they recommended Zimmer's licensure be delayed until the NRC better understood what went wrong at Three Mile Island. This approach seemed to be what the communities around Zimmer wanted. Aside from the formation of the Zimmer Area Citizens group, a whopping 85 percent of Moscow residents signed a petition in support of a licensure delay. Some local construction tradesmen also did an about-face, deciding that Zimmer's safety hazards for workers and residents outweighed its high-paying jobs. This was particularly astonishing given that the late 1970s was a time of manufacturing and industrial job loss, not to mention another economic crisis that would tip into a recession by 1980.

All of this shows the local anti-Zimmer movement gathering steam, building off the words and actions of David Fankhauser and his neighbors, and Jerry Springer and his coalition. But if we polled the room at one of the 1979 hearings and asked everyone, "Are you the anti-Zimmer movement?" it's unlikely people would have resoundingly answered yes. Still, looking back with the advantage of hindsight, we can see that this loose network of people and groups was forming something with momentum, as other towns with nuclear power plants were, too, after Three Mile Island. The same gumption that drove the 1972 hearings—that the public had a right to voice its concerns to authorities and be listened to—continued to mobilize people. One mother captured it well when she told government officials at a licensing hearing in 1979, "You may be the more powerful, but you are not the majority anymore." One other theme surfaced, too, bonding people together: the sense that they couldn't trust the NRC. This was an old complaint: in 1972, it was the Atomic Energy Commission that seemed unresponsive; in 1979, the AEC's successor agency didn't seem any better—and in fact, some thought it worse after its aloof response to Three Mile Island. People across Cincinnati and in other communities with nuclear power plants were increasingly questioning the NRC's accountability, especially as it continued to defer to utilities and outsource regulatory tasks to them. And as America endured blunders, lies, and failures by other federal officials in the 1970s, government accountability became something a lot of people wanted.[4]

In the early hours of March 28, 1979, a mechanical error occurred in one of Three Mile Island's two reactors, leading to a pressure buildup in the core. That, in and of itself, wasn't that serious, but then safety systems tripped, inadvertently taking away vital cooling water from the reactor core and causing radioactive coolant to overflow to an auxiliary building. Confused and flustered, operators mistakenly limited the supply of cooling water to the reactor core, making the problem worse. The little water that remained began to boil. Exposed fuel rods reacted with the steam, releasing hydrogen. About half of the core physically melted, releasing radiation in the containment building. Through escaping radioactive coolant, radiation levels in the plant's auxiliary building soared, and radiation started to be released outside the plant.[5]

Located on the Susquehanna River, Three Mile Island sat about ten miles southeast of Pennsylvania's capital and just a few miles south of a small town called Middletown, home to around ten thousand people. As the debacle on March 28 unfolded, executives from Met-Ed (the utility in charge) and NRC officials maintained a cool stance, telling the press that there was no health threat to Middletown residents and others nearby. But contradicting that—and thereby causing confusion for residents—the governor recommended pregnant women and small children within a five-mile radius of the plant evacuate. He gave that order two days after the initial accident, out of concern for radiation levels around the plant. Everyone else nearby was told to stay inside.[6]

One week later, the plant was still venting radiation into the air. Nonetheless, the NRC continued to insist that it was a negligible amount being released and was unharmful to "normally healthy people." Who that category included was unclear. Furthermore, as dissenting scientists later said, no one knew *exactly* how much radiation was coming from the plant: there were just too many variables for the NRC's monitoring equipment to fully account for all radiation in the area. So the extent of the health threat, and how it would show itself over time, was then unknown.

Nonetheless, on April 9, the governor canceled his evacuation, and residents began to return to the area. By the end of the month, the NRC had successfully shut down both reactors. Later, when scientists initiated cleanup work on the site, they determined that the malfunctioning reactor had come within thirty minutes of fully melting down.[7]

The extent of radiation released at Three Mile Island in 1979 is still debated today. The NRC maintained that the public was not exposed to a significant amount of radiation. Its post-accident assessment said, "The projected number of excess fatal cancers due to the accident that could occur over the remaining lifetime of the population within 50 miles is approximately one." And yet today, around 40 years after the accident, the area affected by the radiation plume has two to three times higher rates of cancer than the surrounding areas, which some residents believe to be the result of the accident.[8]

But even accounting for the ambiguity and disagreement over health outcomes, the near meltdown had unambiguous consequences for nuclear power. It revealed the limits of safety systems within the plants, showing that static barriers like the containment structure around the reactor and active systems like the relief valves were not foolproof. It also showed the extent to which utilities, engineers, and the NRC had underestimated the role of human error. And as a result, the accident considerably increased Americans' distrust of the NRC. They began to joke that NRC, the supposed watchdog, actually stood for "Nobody Really Cares."[9]

Following the accident, states and communities with nuclear power plants were on alert. In Ohio, Republican governor James A. Rhodes convened a Nuclear Task Force, headed by the state EPA director, to check "all aspects of the state's role in nuclear plant safety." In addition to Zimmer, Ohio had the Davis-Besse, Perry, and Erie plants along Lake Erie, with units in operation or under construction at the time of Three Mile Island. Ohio officials said they didn't want "another Harrisburg." "We're reassessing everything related to this entire nuclear field," Rhodes told the press. "We've got to go back in and take another look [to give the public] all the assurance that we're going to have a program for their own protection." In early April, the task force went to Davis-Besse. One of its reactors was operable but had been shut down for maintenance prior to Three Mile Island. In assessing this reactor, the task force learned that eighteen months prior, it had had a malfunction in a reactor valve identical to the one that went awry at Three Mile Island, built by the same manufacturer. Nervous, the task force then toured Zimmer a few weeks later. Officials from Ohio's Disaster Services Agency joined them. Rhodes's group came away with the conclusion that the NRC should have a full-time inspector at Zimmer to better guarantee safety.[10]

Bill Gradison, the Republican congressional representative for Ohio's first district (which included Cincinnati), agreed. During an interview with the *Cincinnati Enquirer*, Gradison (who was also a former mayor of Cincinnati) stated that nothing should happen with Zimmer's license until the NRC completely assessed Three Mile Island—and beyond that, the license should be contingent on CG&E having clear evacuation plans and independent monitoring equipment at the plant. The Ohio Department of Health then announced it had plans to install its own radiation monitoring equipment by the end of 1979.[11]

Meanwhile, Cincinnati's city hall was inundated with hundreds of letters. Concerned residents asked their city councilors and mayor—then, Bobbie Sterne—to prevent Three Mile Island from "being repeated at Zimmer," as many put it. At the very least, they implored, Zimmer's licensing hearings should be postponed. "In light of Harrisburg we must postpone the NRC hearings until all becomes clear," one man wrote. Another entreated, "For the safety of our city and its inhabitants, please act on this matter." Many called out city council's responsibility to public health, especially since it was now an intervenor in upcoming licensing hearings. "So long as you give C.G. & E. the green light for Zimmer," one man wrote, "you too are partially responsible for endangering us all, and may be in part responsible for a 3-Mile Island being visited on Cincinnati."[12]

City council acted. Jerry Springer immediately drafted a motion directing city solicitors to submit a request to the NRC to suspend Zimmer's hearings until more was known from Three Mile Island. In response, hundreds of letters of support poured into city hall, prompting all but one member of council to vote for Springer's motion. City solicitors then submitted the request to the NRC, as did the other intervenors in Zimmer's licensing. Back in council chambers, the one person who was displeased with this course of action was Republican councilman Walter Beckjord, the former CG&E executive, who called the decision "a little bit reckless." He worried that, by pressing the NRC for a hearing delay, "we might lose something, specifically adequate monitoring safeguards." Beckjord was referring to how city officials had been negotiating with CG&E for weeks over monitoring equipment at Zimmer. Because the station was outside municipal jurisdiction, city officials had initiated talks with CG&E to get the utility to install additional air and water monitoring devices at the plant for the sake of Cincinnati's public health. Utility representatives had dug in their heels in these meetings, and so Beckjord worried that if the city requested a delay in Zimmer's hearings, CG&E would back out of any negotiations over monitoring equipment. It was not an unreasonable concern, really, given CG&E's record of dragging its feet whenever external forces threatened to impede internal management. It didn't matter, though, because Beckjord's Democrat- and Charterite-majority council overruled him.[13]

The intervenors—the city, David Fankhauser, and the Miami Valley Power Project representing Dayton Power and Light customers—were successful. The NRC consented to postpone licensing hearings for Zimmer. But the delay was only for a few weeks. Furthermore, it wasn't *really* a result of the intervenors' petitioning. Rather, it was the commencement of federal investigations, both a presidential commission and an NRC review, to understand what had gone wrong at Three Mile Island and how to prevent it at other plants. Since the NRC anticipated that final reports wouldn't be ready for months, it temporarily suspended the issuing of nuclear power plant licenses, and in many places, like Cincinnati, it paused licensing hearings.

Both reports (which were not ready until October of that year) acknowledged the reality of human error in nuclear power plant operations and explained that "fail-proof systems" could, in fact, fail. For the NRC, this was a rare moment of humility. The reports diverged around *who* was responsible for preventing that human error. While Jimmy Carter's commission criticized the utility running Three Mile Island, it also objected to the NRC's policy of outsourcing quality assurance to utilities, where it became a utility's responsibility to have thorough documentation of proper equipment installation and maintenance and where it was up to utilities to perform and coordinate most inspections for design and operational verification—not the NRC. Indeed, the NRC only performed occasional site visits, and throughout the 1970s, as more nuclear power plants were being built, the commission spent its time on responsive, rather than preemptive, safety checks. Staff responded to specific problems related to plant design and safety, in the form of convening study groups and public hearings, but the NRC did not systematically collect and evaluate data on all operating reactors for safety purposes. The presidential commission attacked this "hands-off" regulatory approach, arguing that such standards led to hazy lines of authority among the government, utilities, and contractors in verifying a plant's safety. The NRC's task force disagreed. It insisted that "the basic responsibility for public safety is to remain in the private sector, in the hands of the individual licensees for commercial nuclear power plants." "We believe," it said, "that it is neither feasible nor practical for the [NRC] staff to review every element of every design." As such, quality assurance remained, post–Three Mile Island, under the purview of utilities.[14]

An aspect of nuclear power that *was* reformed was emergency planning protocol. To get a final license, a utility had to present its plans for accident preparedness. CG&E had already prepared emergency plans, and the NRC and Ohio's Disaster Services Agency had approved them in early 1979. But the government review of Three Mile Island rendered these void. Prior to the accident, utilities only had to consider the low population zone (LPZ) around a plant in emergency planning schemes, a term that was vaguely defined by the NRC depending on the nuclear power plant in question. After Three Mile Island, the NRC tightened up its LPZ rules. It mandated that a utility and local authorities had to plan for the sheltering and/or evacuation of people within a ten-mile zone around a nuclear power plant and for food contamination within the fifty miles surrounding a plant.[15]

As these new government mandates crystalized, forcing CG&E executives to update evacuation plans, they worried about the costs of regulatory delays. Zimmer was by then $650 million over budget, and on top of that, another oil shortage and ongoing inflation had only worsened CG&E's financial predicament since the 1973–1974 recession. Further delays to Zimmer's start date were unwelcome, to say the least, so it was with great frustration that company leaders told shareholders, "The scheduled commercial operation date of the Wm. H. Zimmer Nuclear Power Station was once again delayed. It is now scheduled for operation in 1981 [instead of 1980]." Nonetheless, CG&E officials continued to defend their choice to build a nuclear power plant—so much so that after Three Mile Island, they took out an entire page in the *Cincinnati Post* advocating their position. The utility wrote that "the sequence of events that occurred at Three Mile Island could not occur at Zimmer." Here, company representatives were referring to how Zimmer had a boiling water reactor and Three Mile Island a pressurized water one. It was true that the ways these reactors functioned were slightly different, although CG&E failed to note that the fundamentals were basically the same. Both used water and fission to produce power. On top of that, CG&E didn't dwell on the fact that there could be an issue with a boiling water reactor. Instead, the company simply stated, "In [the event of an accident], Zimmer's highly-trained staff would immediately take the most prudent course of action." The ad was a public relations stunt—but it also wasn't: CG&E executives continued to believe in the merits of nuclear power and that great technological innovation entailed some amount of risk. About two weeks after Three Mile Island, the director of CG&E's electrical production department told the *Enquirer*, in what he saw as a reasonable metaphor, that he felt sorry for the Wright brothers, the aviation pioneers: "They would never get anywhere either today."[16]

But here, CG&E executives were refusing to admit that different technologies carried different risks, some grosser than others. Three Mile Island had convinced more Americans that nuclear power's risks were not worth its potential benefits. And since utilities and the NRC were unwilling to acknowledge this perspective, people became angry at them, as much as at the technology itself. The week after the accident, nationwide antinuclear protests involving thousands of people broke out in Washington, DC; Los Angeles; Seattle; Phoenix; Philadelphia; San Francisco; and other cities. In Harrisburg, Pennsylvania, around 1,000 people from the communities around Three Mile Island descended on the capitol building, demanding the plant's permanent closure. A little over a month after the accident, upward of 125,000 gathered in Washington, DC, to oppose Three Mile Island and other nuclear power plants. In September, another protest drew 200,000 to New York City, one of the largest antinuclear protests still to date.[17]

There were protests in Cincinnati too. Three days after the accident, Citizens against a Radioactive Environment—CARE—staged a twenty-four-hour vigil at CG&E's headquarters in downtown Cincinnati. Members held oversized cloth banners that asked, "Zimmer: Will it be another Harrisburg?" Polly Brokaw, coleader of the group, told the press, "Serious accidents are inevitable. Like all nuclear plants, Zimmer is an accident waiting to happen. It should never open and we are committed to prevent its opening."[18]

Two months later, on Sunday, June 3, 1979, one thousand people gathered at a small riverside park about two miles from Zimmer and marched along the sides of State Route 52 to the power plant. CARE was there, as were many people who lived near Zimmer. Fankhauser attended with his family. The action, sponsored by a coalition of antinuclear groups in central and southwest Ohio, was part of an international weekend of nuclear power protest after Three Mile Island. Seventy miles west and downstream from Cincinnati, in Madison, Indiana, around three hundred people protested outside the Marble Hill nuclear power plant site, then also under construction.[19]

The march by Zimmer snaked from the park to the power plant and then to Moscow Elementary School. People carried signs adorned with human skeletons and others that read "Stop Zimmer" or "Save our planet." Tom Carpenter, CARE's other leader, told the crowd, "This is for the children." From the elementary school, people marched back to Zimmer. With hundreds lining the driveway and others shouting encouragement on the side of the highway, twenty-seven individuals sat cross-legged outside of Zimmer's gate, on CG&E property. The Clermont County sheriff warned them to move—once, twice, and then for a third time. Seventy police in riot gear stood nearby at the ready. The twenty-seven, Carpenter included, did not budge, and deputies began arresting people, taking them to the Clermont County jail on trespassing charges. Seventy miles away, eighty-nine were arrested outside of the Marble Hill plant for scaling a fence there. Kentucky author and farmer Wendell Berry—a soft-spoken man who wrote about caring for the natural world—was one of them.[20]

On arrest, Carpenter and another CARE member, Kim Surber, pled not guilty and requested a jury trial. Some months later, a jury heard their case. Defended by a local attorney, Andrew Dennison, Carpenter told the judge, "I feel I had a right to be there. I feel I have a right not to be a victim of radiation. I felt it was my duty to go there and be heard." The hearing ended in a mistrial after the jury deliberated unsuccessfully for hours. Indicative of the effects of Three Mile Island, one jurist said he was confused about nuclear power and therefore felt conflicted about convicting Carpenter and Surber. He had just watched *The China Syndrome*, a Jane Fonda blockbuster movie about a nuclear power plant meltdown, eerily released just twelve days prior to Three Mile Island. The movie had caused him to ponder if such a thing could happen at Zimmer.[21]

Tom Carpenter became an expert on nuclear weapons production, decommissioning, and remediation. For almost forty years, from 1985 to 2022, he provided legal representation to whistleblowers at the Hanford Nuclear Site in Washington state, the facility that produced the plutonium for the bomb that leveled Nagasaki to end World War II. Functional until the late 1980s, it was one of the most polluted places on our planet and produced more whistleblowers than any other place in America. In 2007, Carpenter founded the Hanford Challenge, a nonprofit watchdog, to track the site's mothballing. About thirty years before that, he became a key figure in the fight against Zimmer, just before Three Mile Island, through his Citizens against a Radioactive Environment group.[22]

Carpenter began this path as a twenty-one-year-old in the spring of 1978. He was working for the Cincinnati Food Co-op in Clifton, a leafy neighborhood north of downtown Cincinnati, near the University of Cincinnati. As head of co-op member education, he was tasked with bringing in lecturers on environmental topics. This suited him. Ever since he had read Rachel Caron's *Silent Spring*, he considered himself an environmentalist and an activist. But, born in 1957, he was sad that by virtue of his young age, he had missed the street protests of the 1960s.[23]

One of the speakers Carpenter recruited for the co-op spoke on nuclear power. He explained what an accident could look like and how, even in normal operations, it would emit low-level radiation. Carpenter thought of Zimmer, being built nearby. The lecturer shifted to nuclear weapons, how they were being developed for the Cold War—which the nightly news told Americans was "reviving." From the late 1960s to the mid-1970s, US and Soviet leaders signed antiproliferation treaties, curtailing conventional and nuclear weapons. People called it *détente*, French for *release from tension*. But by the time Carpenter recruited his guest lecturer, cooperation had broken down. Both sides premiered technologically advanced nuclear weapons, like the US' neutron bomb, a small hydrogen bomb designed to kill people in a limited radius. In response, antinuclear organizations emerged (or reemerged) across the world. Some, like the National Committee for a Sane Nuclear Policy, which had formed in the US in 1957 during another time of heightened American-Soviet conflict, had hundreds of thousands of members. Many disarmament groups also opposed nuclear energy since they believed nuclear power plants, like nuclear weapons production sites, had health consequences for workers and residents. People also noted how both nuclear power and weaponry were created and deployed by federal agencies and departments equally uninterested in civilian input.[24]

Against this unfolding political backdrop, Carpenter decided to attend an upcoming antinuclear demonstration in Colorado. Organized by national peace coalitions, five thousand people gathered on Saturday, April 29, 1978, at the plutonium and uranium component manufacturing Rocky Flats plant near Denver. Daniel Ellsberg—the military analyst who leaked the Pentagon papers to the *New York Times* in 1971 showing US involvement in Vietnam as early as the 1940s—spoke on the dangers of the neutron bomb, which would be manufactured at Rocky Flats. Afterward, the march's organizers held a how-to session for anyone interested in establishing antinuclear movements in their own communities.[25]

Carpenter, energized and armed with information, returned to Cincinnati. He wanted to start a big grassroots antinuclear organization that would use street protest and civil disobedience when needed. He was aware that Cincinnati already had an antinuclear organization, Ohio Valley Citizens Concerned about

Nuclear Pollution, founded in the early 1970s. Elizabeth Seeburg (if you remember, David Fankhauser's elderly neighbor) was a member. But it wasn't a very large or active group. Through the 1970s, it sponsored activities to raise awareness of Zimmer and to encourage, as an alternative, energy conservation, but these were infrequent events, held about once a year. Its executive secretary, though—Polly Brokaw—was a prolific activist. Carpenter turned to her when he got back from Denver, asking her to create a new antinuclear organization with him.[26]

Brokaw was a longtime Quaker pacifist and campaigner for civil rights, peace, and environmentalism, as was her husband, Amos. The two devoted their lives to these causes. They met as students at Ball State Teachers College in Muncie, Indiana, where both were involved in desegregation efforts. During World War II, Amos—as a conscientious objector—quit his government job and refused induction. He did so again in 1949 and again in 1952, serving prison sentences three separate times. When he was in federal prison in Kentucky for refusing to serve in the Korean War, Polly moved to Cincinnati to be near him. There, she (and he, on his release) raised their children (including David Fankhauser, from Polly's first marriage) along with many foster children. The Brokaws helped to desegregate Cincinnati's amusement park, were a fixture at anti–Vietnam War marches in the city, and vocally opposed the Cold War arms race and nuclear energy.[27]

Brokaw accepted Carpenter's invitation, launching CARE (and presumably shutting down the other antinuclear organization since, after the spring of 1978, we don't hear any more from that group). CARE's first meeting—around one year before Three Mile Island—attracted about thirty-five people, a mix of students from the University of Cincinnati and older people, recruited by Brokaw. From there, meeting every two weeks, CARE quickly expanded to include a few hundred people who agreed with Carpenter and Brokaw that nuclear weapons and energy were equally problematic and that the US should instead be investing in energy conservation and the nascent renewable energy industry.[28]

With that as its broad-based platform, CARE focused on shuttering Zimmer as it also organized against other places. The Fernald Feed Materials Production Center, located about twenty miles northwest of Cincinnati, was another target. Since the 1950s, it produced high-purity uranium metals for the federal government's plutonium production reactors. Aside from larger critiques—like why, in the first place, the federal government was making nuclear weapons—Carpenter and Brokaw were concerned that radioactive contamination from Fernald was getting into nearby water systems (which was indeed happening—but that's another story).[29]

To grow membership, Carpenter and Brokaw ran a public education campaign about the dangers of nuclear technology. They hosted guest lectures by scientists and presented their group in a diversity of places—from PTA meetings and rotary clubs to college campuses. They distributed informative leaflets ("Let OPEC Keep Their Oil: America Has Plenty of Energy. . . . Our Shortage Is in Leadership") often in collaboration with other local antinuclear and environmental groups. They also waged a publicity campaign in local papers. In one full page in the *Cincinnati Enquirer*, they called Zimmer "The Most Dangerous Idea to Ever Hit the Midwest."[30]

Carpenter and Brokaw used nonviolent civil disobedience and (at times, playful) direct actions to get their point across. Carpenter in particular liked to (comically) push CG&E's buttons. Once, he and others in CARE built a solar collector, painted CG&E's logo on it, and marched it down to the utility's downtown headquarters. A floor manager immediately came over, upset by the commotion, to tell them to take their contraption and leave. Carpenter simply replied, "No, no, this is just a gift." He deposited the solar panel in the lobby and left CG&E employees, dumbfounded, staring at each other. And in another inspired and rather humorous move, Carpenter purchased a single share of CG&E stock. That enabled him to attend every shareholder meeting from that point on. This was, of course, much to the chagrin of CG&E executives, who quickly came to know the young man.[31]

CARE further aggravated CG&E leaders by picketing outside of company headquarters (they usually did this during stockholder meetings). The group also held demonstrations outside of Zimmer. In this way, they echoed earlier antinuclear activists from the 1950s and 1960s who, inspired by sit-ins of the civil rights movement, occupied atmospheric bomb test sites to "bear witness" against what they saw as a great evil. CARE usually demonstrated beyond CG&E property lines to avoid trespassing charges, although that was not always the rule (as Carpenter's arrest at the June 3, 1979, action underscored). Sometimes, in Carpenter's words, arrest was less of a risk than radiation.[32]

One of their demonstrations was a "radioactive release rally" in the summer of 1978. People gathered near Zimmer, and each person released a balloon with a postcard attached to it. Carpenter had come up with the idea: he was trying to show the local wind pattern, where radioactive particles would travel once Zimmer operated. Prevailing winds in Cincinnati usually traveled from west to east, but those near Moscow moved from the southeast to the northwest—so from Zimmer directly toward downtown Cincinnati. The stunt worked, and postcards were returned to CARE to show how far "the radiation" went.[33]

In another event, on Sunday, October 29, 1978, a clear but chilly day, CARE assembled about 250 people at Cincinnati's downtown plaza, Fountain Square, for a Stop Nuclear Power rally. Participants carried signs—"The Future's Dimmer Because of Zimmer"—and heard from David Fankhauser alongside Robert Pollard, the former NRC engineer-turned-whistleblower who became a nuclear watchdog with the Union of Concerned Scientists.[34]

Five months later, at its first meeting after Three Mile Island, eight hundred people showed up. In that moment, the impact of the accident on the Zimmer story, and CARE as a part of it, is clear. Suddenly, the warnings of David Fankhauser and his neighbors, and of Jerry Springer and his coalition, and of Polly Brokaw and Tom Carpenter seemed prescient, discerning—not over the top. Fankhauser and the others represented a segment of

Americans who, in the 1970s, were affected by environmentalism and civil rights and the reviving Cold War, to the point that they questioned nuclear technology on principle. But for most Cincinnatians, just like for most Americans—since there hadn't been a nuclear power plant accident, just like there hadn't been a nuclear war—it took Three Mile Island to break down their indecision, or apathy, regarding nuclear energy.

From 1979 onward, CARE—and in particular, Carpenter—became a fixture in the anti-Zimmer battle. Yet street protest didn't play a huge role moving forward, in contrast to other local antinuclear campaigns. In the late 1970s and early 1980s, thousands of protesters regularly tried to occupy the Diablo Canyon nuclear power plant in California. At one demonstration in 1981, over ten thousand people participated in a nonviolent blockade. On the other side of the country, it was a routine occurrence for thousands of people to protest at the Seabrook nuclear power plant in New Hampshire. The alliances coordinating these assaults drew from a steady supply of people resembling Polly Brokaw—people who had been in activism since the 1960s, if not long before. It stands to reason that there were fewer people of that mold in a Midwest (borderline southern) city like Cincinnati, especially in its more conservative suburban and rural areas, where controversial or combative direct action might not be appealing to many.[35]

But it was also the case that Zimmer's fate was going to be decided indoors—in a conference room in front of a licensing board. And Tom Carpenter understood that. As hearings were set to resume after Three Mile Island, he and Polly Brokaw made the group an official member of the Miami Valley Power Project since it was an intervenor. That way, CARE could directly influence the licensure process.

Eight weeks after Three Mile Island, on Monday, May 21, 1979, Charles Bechhoefer called to order Zimmer's first licensing hearing. As the chairman of the Atomic Safety and Licensing Board, he had traveled from Washington, DC, to the Potter Stewart US Courthouse in Cincinnati's downtown with the stated goal of reviewing two public reports. Like with construction permits, operational licenses for nuclear power plants required the review of topical reports prepared by NRC staff (which included input from other federal and state agencies). Bechhoefer and his board were ready to discuss the thick binder on Zimmer's operational impact on the environment and another on its safety-related design features. (The other reports, one on CG&E's financial preparedness and one on its accident preparedness, weren't ready, partly because the NRC was still finalizing its post–Three Mile Island protocol for emergency planning.)

Like John B. Farmakides, who headed the 1972 public hearings, Bechhoefer was a man who appreciated order in a courtroom and could be gruff when he didn't get it. He was born in Minnesota in 1933 and went by "Chuck" with family. After beginning his legal career at the US Housing and Home Finance Agency in 1958, he worked as a lawyer for the Atomic Energy Commission before joining the Atomic Safety and Licensing Board panel as an administrative judge in 1978.[36]

In Cincinnati, Bechhoefer's temperament quickly emerged when unprecedented numbers of people crowded into each of the 1979 hearings (which dragged on through November) and, during the time allotted to public statements, veered into generalized critiques of nuclear power. To them, Three Mile Island called into question the entire premise of power production by nuclear fission. "It ain't worth it," one man told the licensing board repeatedly. He and many others present were not technically trained scientists or experienced activists. "I'm not even an environmentalist," one woman at the May 23, 1979, hearing said after she made known her opposition to Zimmer. Their numbers were augmented by people who *were* technically trained scientists and experienced activists, including Fankhauser. And for the first time, people came from all over the tristate region.[37]

Many residents queued in front of Bechhoefer had something to say about Zimmer during normal operations, *not* during an accident. One woman from Western Hills, a Cincinnati neighborhood thirty-five miles from Zimmer, told the board she was concerned about the daily emissions of radiation that would come from the plant and how they would affect her water: "I want to know before a month after it happens that I shouldn't have been drinking the water." Another woman took the podium to say, "I'm worried about my safety in the event that that plant goes online." She had read Sternglass's paper—the one that had caused so much controversy for the city of Cincinnati—and between that and Three Mile Island, she was concerned that Zimmer's operation would increase cancer rates in the area: "I have a history of cancer in my family, and my father died of cancer . . . all of my mother's sisters have died of cancer." And while she said, "I admit the idea of a nuclear plant is very seductive. We do need energy," she followed with the caveat that the benefits associated with nuclear power did not offset its risks, especially when human error was considered: "I think if everything worked out according to plan, to what the scientists had laid out, a nuclear plant would be a fantastic idea. But the point is though that we're dealing with human beings. We make mistakes." The man next in line called nuclear power "Pandora's box." "We cannot open it," he said. He concluded, "There's no need to go through with this technology because there are too many questions involved with it." Another woman had an even pithier finish. She looked Bechhoefer in the eye and said, "Don't play craps with our future."[38]

People also told Bechhoefer that Zimmer, in its normal functioning, would destroy the nearby environment. "If Zimmer opens," one woman (who didn't live near the plant) pled, "our land will be forever contaminated; even if the plant is closed at some later date due to the inevitable demise of nuclear power, this is way too late." One woman (who did live near Zimmer) echoed her, rapid-firing questions at Bechhoefer: "The cooling tower is filled with asbestos baffles. What effect will the asbestos particles have on our health? We raise most of our vegetables. How does radiation affect them and will they be safe to eat? What effect will it have on the beef that we raise to eat?"[39]

Some people took the time to suggest to the Atomic Safety and Licensing Board, and to CG&E and NRC officials who were sitting nearby, that Cincinnati should be investing in energy conservation and the emerging wind and solar energy industries. Many people, though, didn't have a solution for regional energy needs; they just knew they didn't want a nuclear power plant. In this way, people's comments were at times panicked or overstated. One man, who told the board that he lived in downtown Cincinnati, said that Zimmer would lead to the "catastrophic destruction of a lifestyle as we know it." And one woman told the board that after Three Mile Island, she and her husband took a drive to see Zimmer: "And all of a sudden, there it was, the cooling tower. And the first thing that struck me about the cooling tower was how it resembled an atomic bomb's mushroom stem." Zimmer didn't have anything to do with nuclear weapons, of course, but it's understandable that people were scared. (And post–Three Mile Island, with the concurrent Cold War arms race, she wasn't the only American making such comparisons.)[40]

Unsurprisingly, many people wanted to talk about the possibility of an accident at Zimmer, including that the Cincinnati area wasn't prepared to respond. Even though CG&E was reworking its accident preparedness, residents emphasized that from what they knew of the existing plans, they were inadequate. One man joked that CG&E's only plan seemed to be "divine intervention." In the face of these comments, Bechhoefer continually reminded the public that they should refrain from making statements on such topics until NRC recommendations had crystallized and furthermore that they should stick to the same script as him: discussing *only* the then-available NRC reports.[41]

That did not happen. Instead, a large contingency of parents and educators from around Moscow, fearful of Zimmer's impact on children, detailed the plant's proximity to growing bodies: within a ten-mile radius of Zimmer stood eleven elementary schools, four middle schools, and five high schools. Moscow Elementary School was less than half a mile away. At one hearing in November, a schoolteacher brought her entire class to show them "democracy in action." She stated to the board, "Kids have vivid imaginations, but imagining radiation is hard, almost impossible. When kids can't see things, they usually imagine the worst. That's why children living around Three Mile Island are still suffering from nightmares. . . . And that's why the kids in Moscow, Ohio, are still wondering what's going to come out of that cooling tower." She paused and then concluded, "So I pose these questions to NRC staff: what are your plans for the children of the world? Is it because they have few rights under the law? Is it because they don't provide capital gains for big companies?"[42]

One mother after the next walked to the front podium and faced Bechhoefer's board. "It doesn't look to me like my children are even going to have a chance to grow up in an environmentally safe home the way I did and the way you did," one told them. Another, the mother of two young daughters, stated that "nuclear power is a completely unwarranted attack on the rights of future generations." "It seems obvious to me," she told Bechhoefer, "that we've no right to expose our children and grandchildren to the increasing threat of incurable disease, and I don't think we have any right either to burden future societies with the problem of dealing with the nuclear waste that we created." Another woman, turning to NRC and CG&E representatives, said, "I feel the NRC, the government, and the utilities are sentencing my children to death."[43]

In thinking about their children or students, people were emotional. Yet as they cried and implored, several made the point that their passionate outbursts were the only *appropriate* response to such a situation. "And I know that perhaps we are said to be very emotional people, but it's a very emotional and trying thing," one mother insisted. Another said the same: "These issues are always dismissed as emotional. But the concerns of this community are genuine, and do deal with emotional issues. The safety of our children, the question of health, are emotional issues if anything. Cancer is an emotional thing." Kim Surber—arrested with

Tom Carpenter at the June 3, 1979, action at Zimmer—told the board panel that she had committed civil disobedience at Zimmer with one person in mind: her nine-year-old son. She was terrified for his safety if Zimmer became operational because she did not trust the NRC's "rubberstamp procedure of licensing." The commission was, in her words, "the epitome of irresponsibility." Tom Carpenter, while not then a father, used similarly passionate language, calling on the government officials in front of him to see the "higher law, called morality. . . . You cannot, in all good conscience, tell me that you think nuclear power is safe. You cannot, in all good conscience, tell me that you can store nuclear waste for hundreds of thousands of years safely, that, in all good conscience, you don't think radiation causes cancer, especially when your own Nuclear Regulatory Commission has said that there is no safe level of radiation." A mother followed him. She said that anyone who told her that radiation was safe "lacks integrity because they know perfectly well there is danger in irradiation."[44]

In these statements, Zimmer's 1979 hearings sounded like the 1972 ones, where parents stood up on behalf of their children. By the late 1970s, more national crises connected to the health and safety of children, infants, and pregnant women had occurred, and these events formed a grim but motivating backdrop to Zimmer's 1979 hearings. Love Canal in upstate New York was one of the worst. In the mid-1970s, parents there discovered that their neighborhood sat on top of a chemical dump. Mothers and fathers mapped their families' illnesses—a high number of birth defects, stillbirths, miscarriages, and rare childhood diseases—to the site. This became typical of the US in the late 1970s and early 1980s. Hundreds of groundwater contamination cases emerged in different towns across the country, leading one EPA official to comment, "The more we look, the more we find." In Woburn, Massachusetts, during the same month as Zimmer's first licensing hearing in 1979, state authorities discovered that industrial solvents trichloroethylene and perchloroethylene had been leaking into the local river, contaminating people's drinking water. Parents agonizingly realized that that most likely explained their town's extremely high rate of childhood leukemia. In response to the toxicity that seemed to be everywhere, Congress passed the Toxic Substances Control Act in 1976 (giving the EPA the authority to regulate and limit chemical substances like PCBs, asbestos, radon, and lead-based paint) and the Superfund legislation in 1980 (to coordinate the remediation of toxic waste sites).[45]

In Love Canal, Woburn, and many other places, parents didn't just grieve. They mobilized, demanding state and federal remediation and relocation, launching lawsuits, and making health-focused parental activism a fixture in American society. Three Mile Island parents did this too. When their utility company tried to restart the reactor that hadn't malfunctioned, and when the NRC convened hearings on it, parents crowded into the room, shouting no. They presented their own evidence of illness (stories of miscarriages, sore throats, nausea, vomiting, diarrhea, misshaped plants, animals with missing parts, large numbers of dead fish floating on the Susquehanna River, all of which occurred after the accident) to show the NRC and the utility that they had underestimated the extent of the damage in the first place and shouldn't be restarting anything. (This was unsuccessful; the reactor resumed operation in 1985.)[46]

It stands to reason that mothers and fathers in Cincinnati (and others who cared about children) were encouraged or at least informed by this parental activism—even if the big difference was that in Cincinnati, no accident had yet occurred. (To this point, in other communities with nuclear power plants under construction, parents also formed an important corps of the local protest movement. The group Mothers for Peace became instrumental in opposing the Diablo Canyon power plant.)[47]

In the midst of many voices at Zimmer's 1979 hearings, one woman, Alice Gerdeman, became a key spokesperson for children's safety. She didn't have any children herself, but she was the principal of an elementary school in Newport, Kentucky, about thirty miles downstream from Zimmer. By the time of the hearings, she had been an educator for almost fifteen years, seven of them in administration. She knew what it took to run a school well, and she cared deeply for the physical and emotional welfare of children. “There does not seem to be any evidence that radiation is good for people,” she told Bechhoefer’s board on Wednesday, November 14, 1979, “and there also does not seem to be any evidence that it is possible to run a nuclear power plant without increasing radiation in the area.” Gerdeman pointed out that in the event of an accident, any sudden release of radiation would impact children’s developing bodies and cause immense emotional trauma.[48]

She was also a nun and had significant activist experience. Born into a Roman Catholic family in the tiny town of Kalida, Ohio, the oldest of eight children, Gerdeman was raised by devout parents—farmers—who emphasized social justice at home; they hired migrant workers each year and taught Alice the importance of treating them equitably. After grade school, she went to boarding school at the Sisters of Divine Providence in northern Kentucky and at the end of high school joined their order. She then earned a bachelor’s degree in elementary education and two master’s degrees, one in education administration and another in theology. From there, beginning in 1965, she taught in Cincinnati area schools staffed by her religious community. And after years of self-imposed training and workshops, she also became head of peace and justice activism for the Sisters. In that capacity, she researched social justice issues, prayed about them, and then took stands on them on behalf of her order. From there, she led her Sisters to street protests, to community meetings, and, with her educational background, to schools. She introduced nonviolence curriculum to parochial schools; she worked with nurses and cooks on justice issues relating to nutrition inequalities and unfair trade practices. The issues she—and thus her order—cared about were wide ranging: they included the death penalty (against), immigration liberalization (for), the Vietnam War (against), and nuclear power (against).[49]

And while Gerdeman was small in stature and soft-spoken with friends, she could be direct and to the point, confrontational if necessary. That was the stance she took when she spoke to the licensing board and to CG&E representatives. "While I am opposed to all nuclear power," she told them, her clear blue eyes leveled at them, "I am particularly opposed to the Zimmer plant because of its location. Putting a nuclear power plant in an area where the main wind drafts go directly over a highly populated area, seems to me to be an entirely foolish way to handle a situation." She continued, "For the last six months, I have been studying the issue of nuclear power because it has concerned me greatly, being in charge of children each day. I'm particularly concerned about the danger to their health and to my own."[50]

Gerdeman was a leader of the new Zimmer Area Citizens group, formed immediately after Three Mile Island. Made up of around forty people who lived close to Zimmer, mostly parents and educators, ZAC—as it was called—was led by Gerdeman along with Margaret and Gene Erbe, parents of eight children, and Genevieve and Andrew Dennison. Andrew was the attorney who represented Tom Carpenter at his trial for trespassing at Zimmer, and in fact, he was known for representing anyone who, in his opinion, had a good cause—many criminal defense cases, many freedom of speech cases. "All underdogs. All Andrew Dennison's clients," Ben Kaufman at the *Enquirer* wrote about him. A veteran of the Korean War and a lifelong Ohio resident, Dennison received his JD from the University of Cincinnati and was admitted to the Ohio bar in 1964. His wife, Genevieve, was from a tiny town in Appalachia, where Ohio and West Virginia meet. She and her husband lived in New Richmond, Ohio, where they raised their son. And while Andrew provided legal counsel to ZAC, it was his wife who was the group's most frequent spokesperson and organizer.[51]

ZAC had a partner organization, Zimmer Area Citizens-Kentucky, or ZACK, based in Mentor, Kentucky, across the Ohio River from Zimmer. Two high school teachers, Mary and Donald Reder, spearheaded it, and Deborah Webb, an attorney and parent from Mentor, represented the group. Given everybody's proximity to the plant—Gerdeman, for instance, lived at the St. Anne's convent in Melbourne, Kentucky, ten miles downriver from Zimmer—Zimmer Area Citizens foremost sought to protect the families closest to the plant. They didn't, though, advocate moving Zimmer to a different location as a way for them to avoid its safety pitfalls; they didn't want parents and children anywhere to deal with it. Some in ZAC and its sister organization were like Gerdeman, of the opinion that Zimmer should be canceled completely. Still, these groups formed with the preliminary goal of getting CG&E and the NRC to create comprehensive and systematic emergency plans in the event of an accident at Zimmer. They believed this outcome likelier than Zimmer's cancellation, and in this way, Zimmer Area Citizens showed themselves to be what they were: pragmatic working professionals, many with young children, determined to protect them.[52]

They started meeting with city, county, and state officials over evacuation concerns, collecting three thousand signatures of support for a delay in Zimmer's licensure until emergency plans were presented to and approved by the local community. After ZAC and ZACK mailed the petition to the NRC, Margaret Erbe as their spokesperson told the *Cincinnati Enquirer*, "We wish to delay the licensing hearings until all of these local agencies can get their act together for the safety of residents." Her husband, Gene, agreed: "They always put up the traffic lights after someone has been killed. We'd like to change that."[53]

The involvement of people like Margaret and Gene Erbe shows how, in the very communities meant to benefit the most from Zimmer, the tide was slowly turning. When a former mayor of Moscow said at one of the 1979 hearings, "I can speak for 75 or 80 percent of the people in Moscow that they could care less one way or the other. The plant does not concern them," a Moscow resident stood up and challenged him. He held a petition with 85 percent of Moscow residents' names on it, demanding a delay in Zimmer's licensing until the safety of the plant could be better guaranteed. To his point, the Moscow Village Council, the New Richmond Village Council, and Washington Township Trustees—government bodies in the immediate Zimmer area—all wrote in support of a delay in Zimmer's licensure until the causes and lessons of the Three Mile Island accident became clearer.[54]

In perhaps an even more remarkable turnaround, there were some local construction workers who decided that Zimmer's health risks trumped its high pay. They were thinking of themselves as workers, and they were thinking of the families who lived nearby. One man, a steelworker, told Bechhoefer's board that one thousand members of his local union decided to oppose Zimmer after the Three Mile Island accident. He himself then joined Citizens against a Radioactive Environment.[55]

This was a significant change of heart for the rural, economically depressed communities around Zimmer that needed the high-paying, unionized trades jobs it offered. The poor economy of the 1970s, coupled with deindustrialization, hit America's working class hard. After almost doubling since 1960, wages for male workers peaked in 1972, and thereafter their real earnings stagnated and began falling. From 1969 to 1979, Ohio lost 79,000 manufacturing positions in primary metals, electrical equipment, machinery, and transportation equipment. While the area around Zimmer was more agricultural than industrial, it nonetheless felt the downturn, making Zimmer's construction jobs and permanent operator positions all the more imperative for the area. To build the plant, CG&E and its contractors employed around 2,500 construction workers and engineers who moved to or lived in nearby rural communities. $168 million was paid out in wages to them, so it wasn't a hasty decision for the people benefiting from Zimmer to caution against its licensure. They clearly arrived at that decision after considerable thought.[56]

However, it certainly wasn't true that everyone around Moscow opposed or questioned Zimmer after Three Mile Island, just like it wasn't true that all of Cincinnati turned antinuclear overnight. Throughout the 1979 hearings, CG&E's legal counsel,

Troy Connor, a man with little patience, was on a mission to get the licensing hearings concluded as quickly as possible. To expedite a process largely out of his control, he called forth supporters of CG&E and Zimmer. These were largely the same people who had endorsed Zimmer from the beginning: local physicists and engineers who believed in nuclear power's safety; business leaders who wanted Zimmer to power the area's economy; and local construction trade union leaders whose men were employed by Zimmer.[57]

But it is true that Three Mile Island did something to the American psyche when it came to nuclear power, and the suburban and rural towns that were home to the plants did see a significant uptick in antinuclear activism after the accident. There was no better example than Three Mile Island's hometown of Middletown. The accident and the NRC's tight-lipped response to it turned a largely pronuclear town—a town that, like Moscow, had welcomed the jobs and tax benefits—into a largely antinuclear one.[58]

On Wednesday, June 20, 1979, David Fankhauser stood before CG&E representatives, Bechhoefer and his board, and NRC officials, trying to convince them that Zimmer posed a real threat to the 135 schoolchildren at Moscow Elementary. At least Bechhoefer sat forward in his seat, listening. His two other board members were leaning back in their chairs at their front panel desk, looking less than engaged. (One person in the audience even commented, "Mr. Hooper," who was one of the board members, "was dozing off during the testimony.") Meanwhile, the attorney the NRC sent to the hearings, a slight and intense man named Charles Barth, was pacing back and forth, muttering to himself as Fankhauser spoke. Taking long, quick steps, balancing his glasses on one finger, Barth stopped only to say something to his NRC colleagues or to slide over to reporters, where he remarked on the irrelevancy of Fankhauser's comments. He even brought with him his hotel's "do not disturb" door sign, which he propped up by his seat. Nobody had any question as to what mood the man was in.[59]

When Fankhauser finished speaking, Bechhoefer informed him that his testimony would largely be considered "hearsay" since it was gathered from "other sources," presumably meaning those outside the NRC. CG&E representatives then asked a man from Sargent and Lundy, the engineering firm behind Zimmer, to address the deluge of statements about children's safety. The man, assistant head of nuclear safeguards at his firm, stated that students at Moscow Elementary would receive only a tiny fraction of the radiation limit set by the federal government. CG&E officials said there you have it; children would receive *almost* no radiation.[60]

This was exactly the opposite of what parents like Fankhauser were looking for. They wanted government and utility leaders to be transparent and cautious with Zimmer's licensure, not self-assured and dismissive of others. Furthermore, people wanted the NRC and the licensing board to dig into what CG&E was saying, not take it at face value. These issues were readily apparent when officials and intervenors finally got around to discussing the reports on Zimmer's environmental impact and its safety-related design features: Fankhauser and others noticed that the conclusions NRC staff had drawn were heavily reliant on CG&E's own conclusions and that criticism from other federal and state agencies had been largely trivialized.

The environmental report anticipated biodiversity loss (i.e., fish, their larvae and eggs, plankton, and bottom-feeding drift organisms caught in Zimmer's circulating water system would be killed due to thermal and mechanical shock), yet NRC staff said that this "fell within acceptable limits of harm" and was "not expected to constitute a significant impact." The commission emphasized that the area was "an already strongly disturbed area of Ohio" without much biodiversity prior to any work by CG&E. Furthermore, the NRC wrote, the area had naturally high levels of metals and inorganic phosphorus and nitrogen, making "this stretch of the Ohio River . . . a poor medium for propagation of aquatic life." Despite identifying thirty-eight species of fish in the vicinity, the NRC reckoned that the environmental impact of the power station on them would be slight because populations there were already "minimal." (Commission officials even claimed that Zimmer would *encourage* natural preservation in the area—"that the non-excavated portion of the plant site will function ecologically as an informal preserve for the common species of the region provided that no further disturbance takes place"—a statement that plainly didn't consider any impacts of radiation from Zimmer.)[61]

These assessments came together from on-site visits made by NRC staff, but they also relied on CG&E's own vegetation studies, species censuses, and soil descriptions. These showed, according to the commission, that "no unusual or critically important biological factors in the terrestrial environment are jeopardized by this project." The NRC said that CG&E would conduct environmental tests in the future and would engage in "best practices" once Zimmer was operational.[62]

The final safety report also diminished risk. It said Zimmer's radiation releases would be "within the applicable limits." NRC staff had written, "There is reasonable assurance (a) that the activities authorized by the operating license can be conducted without endangering the health and safety of the public. . . . The applicant is technically and financially qualified to engage in the activities authorized by the license." And while they noted that CG&E had "not considered" a "hypothetical sequence of failures more severe than Class 8"—Class 8 included a loss of coolant in the reactor, leading to a possible meltdown—NRC officials insisted that such a sequence of events was uncommon, so "their environmental risk is extremely low." The report concluded, "Defense in depth (multiple physical barriers), quality assurance for design, manufacture and operation, continued surveillance and testing, and conservative design are all applied to provide and maintain a high degree of assurance that potential accidents in this class are, and will remain, sufficiently small in probability."[63]

One problem was that the NRC had finished this report before the Three Mile Island accident and before the government investigations into the accident had concluded. The NRC attorney, Charles Barth, acknowledged this: at the June 20, 1979, hearing, he said that his agency was reexamining its written conclusions about Zimmer in light of the accident but that any changes to its report would not likely affect what was being discussed at Zimmer's hearings.[64]

Fankhauser and his fellow intervenors didn't let this go, though. They repeatedly asked Bechhoefer's board: Why are we discussing the reports if they're not yet final? The NRC had written (regarding Zimmer's General Electric Mark II containment vessel around its reactor), "We are currently conducting our review of the Mark II." Fankhauser asked Bechhoefer about this: Why was his board evaluating a report that hadn't finished assessing a key safety feature? It was also the case that during the summer of 1979, General Electric issued a warning for its reactor, saying it had discovered that specific bracing was required for it to meet NRC standards of earthquake preparedness. The NRC then ordered CG&E officials to rework major components of the cooling system, including the pipes from safety valves to the suppression pool. Fankhauser pressed his point again: Why was the licensing board reviewing Zimmer's safety-related design when it clearly wasn't finalized?[65]

When Bechhoefer asked CG&E representatives to clarify these points, they responded grudgingly, usually in legalese and other jargon. Margaret Erbe of Zimmer Area Citizens described it as "the typical PR lecture with slides and the whole bit. . . . Technical language that hedges around and never answers questions." Many members of the public expressed struggling to understand whatever CG&E's lawyer and other officials said, believing that their choice of words was intentionally employed to confuse and intimidate people. This impression just solidified the long-standing one that CG&E leaders didn't care about public proceedings but instead blamed the hearings—and the public who attended them—for exacerbating an already long review process. The result was tense, antagonistic hearings, with intervenors especially feeling browbeaten and nervous. Just before one of the hearings, people overheard Fankhauser muttering under his breath, "They're going to blow me away."[66]

The upside of this intimidation was that it bonded the intervenors to members of the audience who were critical of Zimmer. People standing in the back of the room often applauded and cheered when an intervenor or member of the public had finished speaking. Bechhoefer, of course, tried to prevent this. He said on one occasion, "I might add that applause is not appropriate for a courtroom." But the man who had the podium challenged him on this: "To me, clapping is a good thing and it supports people as a group, especially in the face of being in a very formal setting like this. . . . We are somewhat intimidated by this." He continued by saying that clapping was helpful when "the public wants to say something but they're not sophisticated enough to say it in the right way."[67]

Emboldened, people criticized not only the NRC and CG&E but also the Atomic Safety and Licensing Board, the entity with the power to recommend or not recommend a license. A young mother commented that nobody from the board lived anywhere near Zimmer: "It's very upsetting for me to sit here and look at you and know that you two men who live so very far away from any danger should have the power to decide anything that would affect the lives of so many innocent people, people who have no voice in the matter at all." And people remarked that as a body, the licensing board didn't look like Cincinnati. A man who lived close to Zimmer told Bechhoefer, "I appeared about seven or eight years ago when the license for construction was asked for, and I notice the composition of the Board hasn't changed. It's still all men, no women, no blacks. . . . But I hope maybe because you are dealing with life-giving problems here that it would be more of a composition of all walks of life." He finished with: "I have a lot of questions about engineers and lawyers making these decisions. I would rather see farmers, musicians, people of this nature who have a bigger outlook on life than I think people who come from such a narrow spectrum."[68]

Disappointed with all the different bodies involved in nuclear power plant licensure, people at Zimmer's hearings were deeply pessimistic of the entire process. One woman said, "The day that I heard that 115 tons of radioactive uranium fuel was being shipped to Zimmer, I called the NRC. . . . On that day I talked to Mr. Charles Barth, the NRC attorney. He stated that Zimmer would be licensed to operate in the spring of next year. I realized what I had already known: that this had been the way for all nuclear installations in the country. Of course fuel is shipped because of course the plant would be licensed. Even after Three Mile Island it appears to be business as usual for the NRC." Another man echoed her: "Because the Board is omnipotent, it has all the decisions to make and we can do nothing but come here and say how we feel. But it's very disappointing that we probably have no say at all." Another woman said the same: "The NRC was to be the watchdog for public safety. . . . Upon examination, it seems—it comes out that most of the people on your Commission have in fact close ties with the industry. It also comes to light that you have never denied an application to operate." People literally damned the licensing board and NRC officials. One person's "I don't know how you can live with yourselves" was followed by another's "In conclusion, gentlemen, I will leave you with something: if you license this plant I wish you the same fate to which you are condemning us, that is, a slow cancerous death."[69]

The vitriol kept coming. A physician and professor of medicine at the University of Cincinnati leveled at Charles Barth that the NRC was meeting only the "lowest common denominator" of safety. A woman, a farm owner in Washington Township, two miles from Zimmer, said to Barth and Bechhoefer, "We have been asked to accept the risk, the health risk to ourselves and our children. . . . We have been repeatedly told that nuclear power is a safe, reliable and economic means of producing electricity. We have been asked to rely on the experts for our safety and well-being." But the NRC, she said, had created a reputation of mistruths that did not make her feel its officials were trustworthy: "We have been told that spent fuel rods would be recycled. We find out there are no reprocessing plants. . . . We have been told that the wastes will be permanently stored, but there has been no permanent storage made available in the 30 years of the industry. We were assured in 1976, February, that adequate monitoring and evacuation plans were being made. To date we have seen nothing." She concluded, "I don't know how you can expect our trust and reliance when there are still so many unanswered questions." Her husband followed her, saying, "I think the nuclear industry requires an awful lot of faith with the public, and I don't, frankly, have that faith."[70]

People's lack of faith in the NRC was specific to its officials, but that distrust developed in an era when the average American had begun to lose faith in many kinds of government leaders and other authority figures, including trained experts. In the 1970s, national hardships and disgraces pummeled the country, showing elected government officials acting incompetently, ineffectively, or dishonestly. In 1965, 75 percent of Americans believed they could trust the federal government. By the late 1970s, that had dropped to 25 percent.[71]

To many Americans, the 1970s felt like a letdown because it came after years of grand promises. Starting in the 1930s, the federal government regulated and intervened in the economy like never before, all in an effort to better serve the public and improve people's quality of life. Officials worked closely with labor organizations and large businesses to do so, as the collaboration between the Atomic Energy Commission and private utility companies—in the name of developing cheap energy for American homes—reflected. Many citizens trusted this system, although in the 1960s, movements like civil rights and environmentalism critiqued federal bureaucracy (particularly the executive branch) as all-powerful and impersonal. They asked, what say does the *public* actually have in government decisions? The AEC, after all, didn't *really* consider public input when it came to nuclear power. Consequentially, 1960s presidents gave community councils and other public forums power to shape the antipoverty, antipollution, and antidiscrimination laws and programs coming out of the White House. While it was mostly the progressive wing of the Democratic Party pushing for these changes, Nixon (a Republican) also continued that legacy to an extent.[72]

But as the 1970s wore on, many people became frustrated: 1960s social programs didn't work as well as intended, disheartening progressive Americans. At the same time, those programs alienated previously loyal Democrats—like white working-class Americans who felt that the party had prioritized the needs of Black Americans and poor families over them. Government coverups in the 1970s over the Vietnam War and Nixon's Watergate scandal further sunk public trust, as did the decade's oil and gasoline shortages and economic recessions. Even Republican Ronald Reagan, elected president in 1980 on the promise of making the federal government as small as possible, couldn't stem the tide of growing distrust in federal leaders. By 1980, fewer people voted then than they had ten years prior. Voter apathy and cynicism spilled into other major institutions, making Americans skeptical of experts in the medical profession, unions, education, banks, the press, churches, and big business, including power companies like CG&E. Environmental disasters further hit a nerve: people in Love Canal and other contaminated communities asked, how could federal and state leaders have let this happen?[73]

In the midst of all this, Three Mile Island became yet another catastrophe in a long line of catastrophes that contributed to a widening credibility gap between ordinary Americans and their leaders. People not only questioned the NRC's leadership but also the official scientific stances its officers took. That distrust, as Zimmer's 1979 hearings showed, galvanized local antinuclear movements in Cincinnati and elsewhere.[74]

At the beginning of 1979, CG&E executives anticipated that all of Zimmer's licensing hearings would be finished by the end of the year and that they would be off and running with further construction. That did not happen. Instead, at the close of 1979, CG&E

was still working to revise its accident preparedness plans for Zimmer and still had yet to present them to a licensing board, requiring more hearings. On top of that, CG&E's financial capacity to finish, operate, and (eventually) decommission Zimmer had also yet to be discussed at Atomic Safety and Licensing Board sessions. So as calendars turned to 1980 and as Zimmer sat there—93 percent complete—the NRC paused licensing hearings, giving CG&E time to prepare more materials.

There was, however, another reason why licensure eluded CG&E. A growing number of workers and ex-workers at Zimmer, with stories of serious defects, began to come forward as whistleblowers.

CG&E officers, 1975. Seated in the center is William H. Dickhoner, a longtime company executive and major player in the battle over Zimmer. Standing second from right is Earl A. Borgmann, another seasoned executive with a significant role in promoting the Zimmer power plant. Courtesy of Duke Energy Corporate Library and Archives.

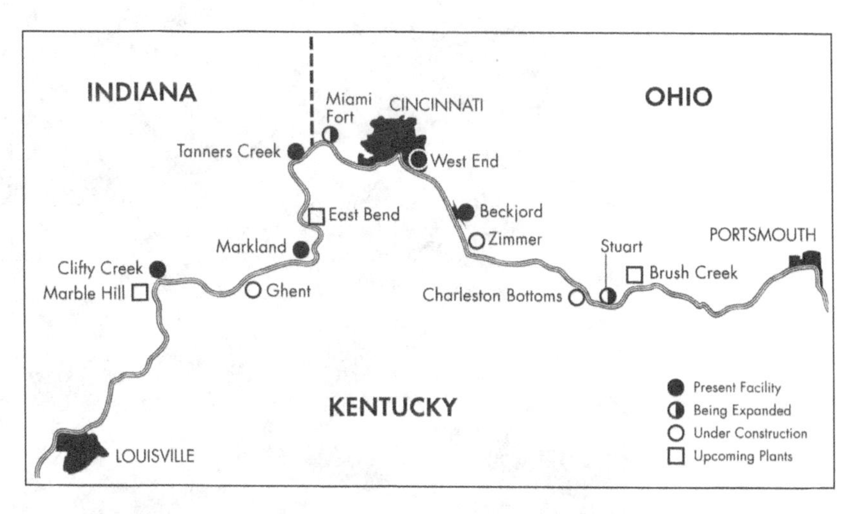

Map of Cincinnati area showing existing and proposed power plants along the Ohio River in 1973—including Zimmer to the southeast of the city. Courtesy of the *Cincinnati Post*, The E. W. Scripps Company. Redrawn by Derek Scacchetti.

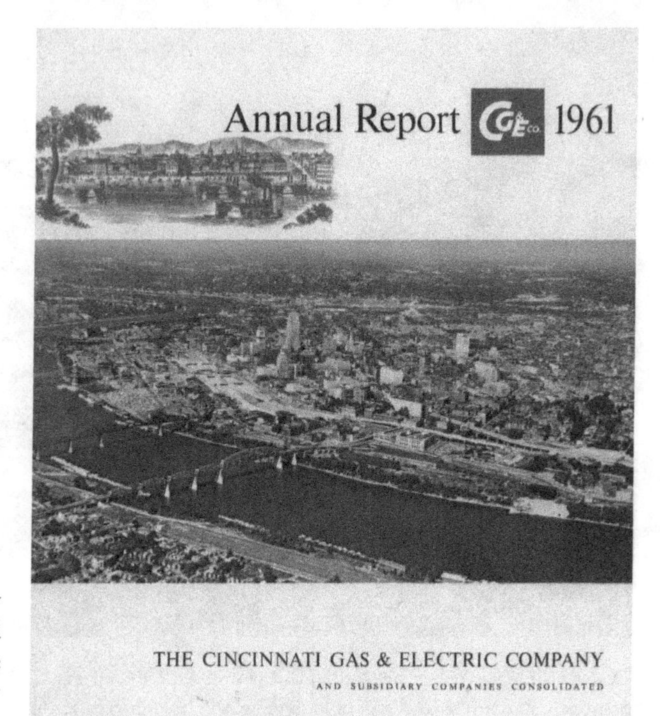

Cincinnati, 1961. A city transformed by outward sprawl. Courtesy of Duke Energy Corporate Library and Archives.

CG&E executives touring Zimmer, 1973. Courtesy of Duke Energy Corporate Library and Archives.

David Fankhauser, one of the longest-standing critics of the Zimmer nuclear power plant, and his family at their home in New Richmond, Ohio. Courtesy of the *Cincinnati Enquirer*.

Troy Connor, CG&E's attorney at the Zimmer permitting and licensing hearings. Courtesy of the *Cincinnati Enquirer*.

The *Cincinnati Enquirer*'s aerial view of Zimmer under construction, 1979. Courtesy of the *Cincinnati Enquirer*.

Jerry Springer—city of Cincinnati councilor and critic of the Zimmer nuclear power plant—on site in Moscow, Ohio. Courtesy of the *Cincinnati Enquirer*.

Ted Berry, Cincinnati's first Black mayor—and someone who consistently challenged CG&E. Courtesy of the Archives and Rare Books Library, University of Cincinnati.

Bobbie Sterne, Cincinnati's first full-term female mayor—and, like Jerry Springer and Ted Berry, an official who was critical of CG&E and Zimmer. Courtesy of the Archives and Rare Books Library, University of Cincinnati.

Chairman of the Atomic Safety and Licensing Board, Charles Bechhoefer (*left*), talking with another board member, Frank Hooper, during the 1979 licensing hearings. Courtesy of the *Cincinnati Enquirer.*

Police confronting protesters at Zimmer after the Three Mile Island accident. Courtesy of the *Cincinnati Enquirer.*

Protest over Zimmer in the aftermath of Three Mile Island. Courtesy of the *Cincinnati Enquirer.*

Tom Carpenter—one of the major anti-Zimmer activists—seen here with his CG&E stock certificate, 1982. Courtesy of the *Cincinnati Enquirer.*

Alice Gerdeman led Zimmer Area Citizens along with others who lived and worked near Zimmer. A veteran protester, she is seen here lighting candles for a Take Back the Night event. Courtesy of the *Cincinnati Enquirer.*

Children playing with Zimmer in the background. Zimmer Area Citizens organized against the power plant with such children in mind. Courtesy of the *Cincinnati Enquirer.*

Tom Devine, the Government Accountability Project attorney who represented anti-Zimmer groups. Author's collection.

Brewster Rhoads, pictured here with his wife, spearheaded the Ohio Public Interest Campaign's work against Zimmer's rising costs. Courtesy of Brewster Rhoads.

Ohio Public Interest Campaign lead organizer, Roxanne Qualls. Here, she is at the University of Cincinnati, giving a speech for Earth Day. Courtesy of the Archives and Rare Books Library, University of Cincinnati.

An aerial of Zimmer in 1982. Courtesy of Grandview Heights Public Library.

The complexity of a nuclear power plant on display: seen here is the area directly beneath Zimmer's reactor containment vessel. Courtesy of Grandview Heights Public Library.

Three Zimmer workers on site, summer of 1982. Courtesy of Grandview Heights Public Library.

Workers by Zimmer's main electrical generators, 1981. Courtesy of Grandview Heights Public Library.

Graffiti inside Zimmer, darkly mocking the nuclear—and shoddy—nature of the plant. Courtesy of Grandview Heights Public Library.

CG&E officials tried to promote goodwill in the Cincinnati community by offering guided tours of Zimmer. Here is one in November 1981, showing the control room. Courtesy of Grandview Heights Public Library.

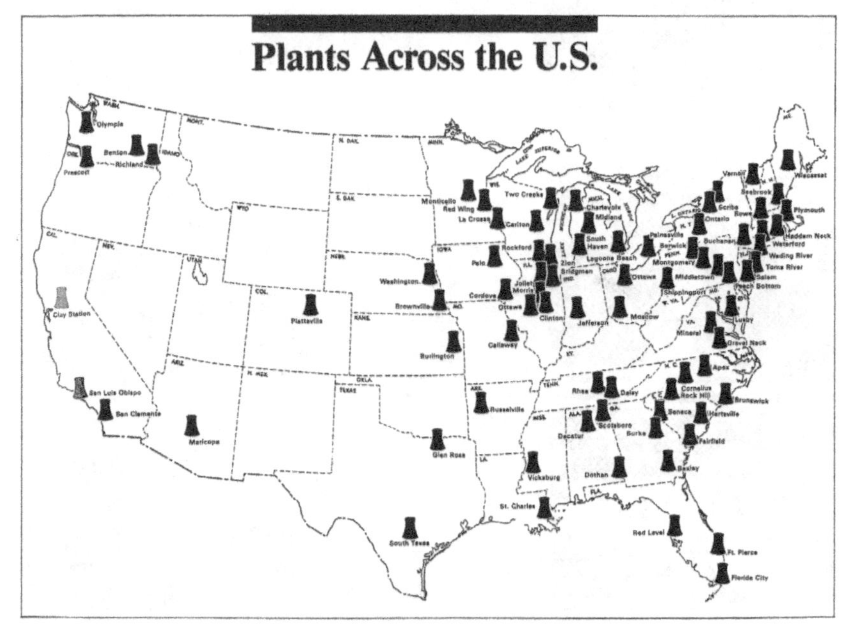

Planned, under construction, and in-operation nuclear power plants across the US, 1983. Courtesy of *Cincinnati Magazine*.

Marian Spencer, playing with a group of children. A lifelong civil rights activist, she was also key to diversifying the opposition to the Zimmer nuclear power plant. Courtesy of the Archives and Rare Books Library, University of Cincinnati.

One of many Nuclear Regulatory Commission licensing hearings. Here, city of Cincinnati councilor Guy Guckenberger questions Nuclear Regulatory Commission officials. Courtesy of the *Cincinnati Enquirer*.

Another public meeting over Zimmer, this time about CG&E's disaster response abilities. Courtesy of Grandview Heights Public Library.

Zimmer in 2017 as a coal power plant. The station permanently shut down five years later. Courtesy of Wikipedia.

5

"The Dirtiest Plant I've Ever Seen"

Zimmer's Whistleblowers

In December 1979, CG&E hired a private inspector, a man named Thomas Applegate, to investigate time theft among employees at Zimmer. He was known to be a bit of an eccentric. Described by those who spent time with him as a "Barnum and Bailey" type of person, Applegate liked to wear a cowboy hat and frequently spoke his mind. He had a legal background in the sense that he clerked at his father's law office in Columbus, Ohio, in the early 1970s. He attended both Otterbein College and Xavier University without graduating and then moved into private investigating. His first job was with Confidential Services Incorporated. Working on a case for them, he happened across a tape of conversations, recorded at Zimmer, showing that workers were illegally selling firearms there. Applegate took the tape to CG&E officials, assuming they would be concerned. The utility ended up hiring him—not to verify criminal activity but instead to see if workers were cheating on time cards.[1]

Applegate accepted the undercover post, posing at Zimmer as a cost accounting engineer for six weeks. Roaming freely at the plant, Applegate observed construction; spoke with workers, including union leaders; and had access to records from the general contractor. While Applegate did note time theft, he spent less time spying on the workers and more time listening to them. Men shared with Applegate serious problems at the plant, including the presence of illegal firearms; drunkenness on the job site; theft of construction materials; inadequate fire detection; poor security measures; installation and operational faults in key safety systems; poor welding work, including cracks in safety welds, which had been covered up; and an ineffective quality assurance program.[2]

Stunned, Applegate took these issues to CG&E, sharing them with Earl A. Borgmann, the utility's senior vice president. Borgmann told Applegate that he appreciated the proof of time theft—but he summarily ignored the rest of Applegate's evidence. In fact, CG&E officials asked Applegate to instead find a reason to dismiss one of the companies performing the quality assurance work at Zimmer. Why? Applegate guessed it was because its employees were known to be incredibly meticulous in their work. Applegate refused CG&E's request. In retaliation, Borgmann released Applegate's findings to workers at Zimmer, who—with their jobs and reputations on the line—subsequently began to threaten Applegate.[3]

In January 1980, CG&E fired Applegate. Borgmann said Applegate's work reached the point that he was taping irrelevant matter from workers' talk at bars. "It was 50 percent fact, 50 percent fiction," Borgmann later commented. CG&E also removed the employees who committed time theft (who were the same employees most vocal about CG&E's lack of quality assurance). These workers were Applegate's early "leads," and CG&E knew this since Applegate disclosed their names in reports to utility leadership. CG&E also fired the quality assurance company that Applegate had refused to disparage. After all of this, the private investigator felt he had to turn somewhere. So he telephoned the NRC's headquarters on Friday, February 15, 1980, and sent his evidence to the Federal Bureau of Investigation (FBI) and the US Attorney's Office in Cincinnati. NRC officials reluctantly met with him, and when they did, they concluded that his charges were unsubstantiated and warranted no action on the part of CG&E. Sidelined, Applegate decided to approach Tom Carpenter and his group, Citizens against a Radioactive Environment.[4]

Before contacting Carpenter in person, Applegate broke into CARE's office and went through its documents. He later told Carpenter that he was trying to make sure CARE was the best organization for him to disclose information to. Satisfied, Applegate secured a meeting with Carpenter the following day, where he admitted to trespassing. Applegate brought with him his box of documents from his time as CG&E's private inspector. In it were transcripts of taped conversations with workers showing that, among many other issues, at least 20 percent of Zimmer's steel came from a junkyard.[5]

Carpenter was eager to assist Applegate (and he didn't mind the trespassing), but he was unsure of how to help the whistleblower. He suggested they consult with Critical Mass Energy Project in Washington, DC, a major antinuclear watchdog founded by consumer rights activist Ralph Nader in 1974. So, in June 1980, the two drove to DC. Applegate, exceedingly nervous, brought with him a small handgun, convinced that something awful was going to happen to them along the way. They arrived in DC without event, though—the gun had been unnecessary—but representatives at Critical Mass, while sympathetic, said they were not the best people to aid the situation. Instead, they advised Applegate and Carpenter to go to a nearby law firm, the Government Accountability Project, a nonpartisan public interest practice created in 1977 to represent whistleblowers. The two men did just that, meeting with one of GAP's attorneys, Tom Devine.[6]

While the private investigator struck Devine as a bit bombastic, Devine was convinced that Applegate was telling the truth. He told Carpenter, "We'll take on Applegate's case if we can partner with you. We need a local presence who knows what's going on in Cincinnati." And with that, Devine came to represent Carpenter's CARE as well as the Miami Valley Power Project as intervenors at Zimmer's licensing hearings. As part of this, Devine also became the legal counsel for Applegate—and a growing number of other whistleblowers who worked at Zimmer, including several who had come forward to the NRC years before Applegate had. They, too, had been sidelined, often fired, supposedly for other offenses. Collectively, these men alleged faulty parts, improper documentation, unqualified workmanship, and dismissive management, and yet, despite the fact that the NRC substantiated much of it, that didn't result in any regulatory shift. The NRC merely ordered parts to be reordered, reinstalled, or repaired, and officials continued to rely on CG&E and its team to do the bulk of quality assurance work—that is, until Devine came to Cincinnati. Throughout 1980 and early 1981, he stitched together whistleblowers' affidavits and subpoenaed government documents to show that Zimmer's components were so compromised and its quality assurance program so lacking that construction needed to stop. From there, he forced the NRC into corrective action.[7]

Zimmer, in its gross number of mistakes, was shocking, but in the years around 1980, the NRC itself received reports of thousands of design and mechanical failures and human errors at licensed nuclear power plants—3,804 incidents in 1980 alone. This revealed how utilities building and operating nuclear power plants in the 1970s and 1980s, and the engineers designing and manufacturing the reactors, shared the experience of inexperience. Beyond that, utilities—dealing with inflation and other tough economic factors—prioritized production timetables for their nuclear power stations instead of ensuring their safety. Compounding these issues was the Atomic Energy Commission and the Nuclear Regulatory Commission's hands-off regulatory style. For all these reasons and others, it wasn't just Zimmer producing whistleblowers. And moreover, it wasn't just the nuclear power industry producing them. Whistleblowing, while not new or unique to the 1970s, was a growing strategy in many fields for keeping government and businesses accountable in the era of Watergate and other national scandals that revealed inept and dishonest leadership.[8]

The impression that the NRC could not be trusted as a regulator had hit the Cincinnati area with the Three Mile Island accident, prompting people to ask, "Could that happen here?" Whistleblowers, risking their jobs and reputations to disclose issues at Zimmer, turned the question into a statement for a growing number. In this way, whistleblowers specified the local battle over Zimmer to be much more *about Zimmer*: there was something specifically wrong with it. Devine's collaboration with these workers was crucial to this transformation, and as a result, local distrust toward CG&E and the NRC swelled. Tom Carpenter understood this. "People became anti-nuclear not from evacuation plans and warnings," he said. "Whistleblowers convinced them."[9]

Created in 1977 as a part of the progressive think tank, the Institute for Policy Studies, the Government Accountability Project was (and is) a nonprofit, nonpartisan public interest law firm specializing in representing whistleblowers against corruption in corporations, nonprofits, and the government. The creation of GAP underscored how, in the 1960s and 1970s, progressive-minded reformers became frustrated with Republican *and* Democrat leadership, prompting the rise of a public interest movement. It was spearheaded by Ralph Nader and others who believed in federal regulation but felt that government officials, entrenched too long in office, needed a watchdog. Like Nader, many public interest lawyers were young, fresh from law schools, and zealous about improving the US. (Additionally, many had been influenced by their own time spent with civil rights and antipoverty activists in the 1960s.) In the 1970s, they took DC by storm, establishing law firms that lobbied for regulatory legislation, litigating against government agencies, and intervening in their administrative proceedings. GAP was one of these firms.[10]

When Tom Carpenter met Tom Devine that summer day in 1980, Carpenter encountered a man not much older than himself. The GAP attorney, who was from Chicago, was not yet thirty. But his attitude and legal prowess were those of a seasoned lawyer who took his work very seriously. He always wore large-rimmed glasses, peering through them very intently. The son of a quality control inspector, Devine was raised to appreciate processes of quality assurance and product safety—and to value the right of an individual to challenge abuses of large businesses and organizations. He attended Georgetown University in the early 1970s and left with a strong desire to change the world for the better. After graduating, he vacillated between being an investigative journalist or a public interest lawyer and ultimately chose the latter. He then went to Antioch Law School, where part of his training involved working as a writer and research assistant for a public interest law firm in Washington, DC. His work there included covering Nixon's attempts to purge Democrats from the executive branch during the Watergate scandal. When the lawyer he was working under was called to speak at an early whistleblower conference in the wake of Watergate, Devine joined them. After sitting through one session, he decided he wanted to craft a law career representing people like Thomas Applegate since, in Devine's words, "wherever there's power, it gets abused." Back at Antioch, he established a whistleblowers' clinic and after graduating was hired in the late 1970s at GAP, where he quickly began to represent whistleblowers across different industries.[11]

After Devine agreed to represent Carpenter and Applegate, Carpenter returned to Cincinnati and assembled all the members of CARE. Devine had impressed on him that for GAP to be involved, CARE needed to fully focus on a targeted legal strategy with GAP and the Miami Valley Power Project—meaning no more street protest. Not all CARE members, especially Carpenter's cofounder, Polly Brokaw, felt that direction was a good idea. The meeting was tense, and Brokaw was resolute that CARE should not agree to discontinue future civil disobedience. Ultimately, though, the majority of members agreed to support the shift. They changed their name from Citizens against a Radioactive Environment to the Cincinnati Alliance for Responsible Energy.[12]

With CARE officially on board, Devine came to Cincinnati and held a press conference downtown, letting everyone know that his law firm would provide counsel for CARE and Miami Valley as intervenors in Zimmer's licensure. While still based in DC, he began flying to Cincinnati frequently, staying with Carpenter during his visits. There, his personality, his lawyering, quickly became evident: he was obsessive in his drive for technical and legal detail and accuracy, and he did not allow for any gratuitous charges against CG&E or the NRC. Everything had to be documented from multiple sources. One of the anti-Zimmer activists later described him as "one of the most serious people I've ever met" who "caught up with nuclear engineers by sheer force of will." Devine was, the activist said, "tireless, someone who slept for three hours a night for months on end to make sure their work was unparalleled."[13]

One of Devine's first moves in Cincinnati was to meet with the whistleblowers Applegate had named. He spoke with a welder—one of the men who had threatened to kill Applegate, actually—who not only gave Devine his statement but also suggested additional whistleblowers who agreed to speak with Devine. From there, other welders came forward. So did quality assurance inspectors and other tradesmen.[14]

As part of his initial evidence gathering, Devine also reached back in time—before Applegate was even involved in Zimmer—to understand the men who came forward to CG&E, to Zimmer's general contractor, Kaiser, and to the NRC with safety-related concerns in the mid- and late 1970s. Coverage of them had filtered into local news and even emerged at 1979 licensing hearings. Devine scoured these sources, getting his bearings on the troubled state of Zimmer.

In 1976, Victor Griffin, a fifty-six-year-old quality assurance inspector at Kaiser, who had been in the industry for twenty years, told his manager that workers at Zimmer were not inspecting materials for defaults prior to installation, nor were they documenting equipment installations thoroughly. This, he said, made his job—to inspect the physical work and approve the paperwork—problematic, and as such he refused to sign off on his inspections in good faith. Kaiser management ignored his concerns, and he resigned—but not before letting the NRC know about his concerns. Officials from the commission responded with an on-site visit, after which they declared everything was fine. One of the inspectors stated that Griffin's allegations were "in essence, true. . . . But it's not a big deal." For one, the NRC had conducted its own on-site inspections—there had been six in 1975—to make sure ongoing construction work was adequate. Furthermore, the NRC stood by its devolved, hands-off style of regulation, arguing that CG&E and Kaiser's off-site paper inspections of equipment complied with NRC standards. The likelihood of any problems during Zimmer's operation was, in the NRC inspector's words, "so remote, it's almost nill [*sic*]."[15]

In 1978, whistleblowers from Husky Products, the company that made the cable trays for Zimmer, came forward. Because cable trays held the electrical cables throughout Zimmer, they were a vital system that needed to be built and installed correctly. One whistleblower told his managers that the trays were being installed at Zimmer by unqualified welders who were using substandard steel. After he voiced his concerns, he was fired. Another Husky employee also came forward with concerns about the cable trays. Though unconnected to the other whistleblower, he raised similar points and was subsequently dismissed from Husky's welder training program. Both men believed they were let go for their whistleblowing. In early 1979, a few months before Three Mile Island, the NRC reviewed Zimmer's cable trays and concluded that they were in adequate condition. But that summer—Devine learned—more men came forward, presenting affidavits to the NRC.[16]

An electrician submitted a statement that cable trays were being overloaded with extra cables. The man feared that by doing so, the generated heat would cause cable deterioration, operational malfunctions, and possibly a fire. He also attested that CG&E and Kaiser used the wrong type of earthquake-resistant cable hangers. "Control devices are in many instances of very poor quality," he wrote in his statement.[17]

A millwright claimed the control rods—the vital parts inserted into the reactor to control the nuclear reaction—were installed without meeting certain specifications and without being inspected by third parties. In his affidavit, he wrote that quality control inspectors "didn't do nothing" to check control rod seals and that his foreman instructed him to "be quiet about the problems at the plant." The whistleblower ignored the advice and told management his concerns. He was then laid off and not rehired for similar work. Another man, an ironworker, disclosed that doors in Zimmer's reactor building and pumphouse failed to meet certain pressure tests.[18]

A quality control inspector maintained in his affidavit that wall plates of the suppression pool—the backup safety system beneath the reactor—had been installed without instruction. He was fired when he tried to raise these issues with CG&E and Kaiser, although CG&E managers maintained that his dismissal was because he had approved defective work. Around the same time, a welder at Zimmer said that metal shavings from poor welding work remained on the blades of the control rods. He explained that the leftover shavings could result in cracks through which boron carbide could "leech out." From that, CG&E could lose control of its reactor if too much escaped. Another whistleblower summarized his feelings about Zimmer: it was the "dirtiest plant I've ever seen."[19]

As Devine familiarized himself with these whistleblowers, he discovered that, following their claims, NRC inspectors traveled to the plant to investigate claims made in affidavits, and some expressed great unease over Zimmer. "This job is very screwed up. Zimmer seems to be the worst case," one NRC official from the Chicago office stated. Through randomized sampling of equipment and systems, government inspectors corroborated many of the whistleblowers' concerns. Officials discovered that supports—holding miles of pipes—lacked proper strength and required an almost total reinstall since workers did not consult architectural plans before installing them. Anti-shock devices for the pipes were unsound, and the engineering design firm for Zimmer never calculated how much strain the supports could hold. Control rods failed to pass inspection. Understaffed construction crews also concerned inspectors. So did the incorrect installation of reactor controls and bent alignment pins in the reactor's fuel bundles.[20]

Despite these serious issues raised by its own inspectors, the NRC simply ordered systems and parts to be reinstalled or fixed. For certain issues—like the overloading of cable trays—NRC officials commented that conditions at Zimmer were "not desirable" but also "not contrary to specifications or NRC requirements." Officials held that some whistleblowers had been overly cautious in their concerns—that some of the improperly installed parts were for *non*safety-related systems and thus were not necessary to repair. In none of its feedback did the commission officially question how and why problems arose in the first place, as that would have meant critiquing Zimmer's quality assurance program and the NRC's own standard of devolved regulation.[21]

As Devine knew, quality assurance for Zimmer started with the plant's parts and systems. The first step was for Kaiser as the general contractor, with CG&E oversight and support, to buy verifiable equipment from manufacturers who could guarantee their laboratories and facilities had fabricated quality parts that met industry safety standards. Then the NRC expected CG&E and Kaiser to hire qualified tradesmen to do the installation—men who would confirm the adequacy of parts as they were being fitted. Finally, inspectors were supposed to document the construction and installation process, using a combination of physical touch, visual appraisal, and computer verification. (It was understood that not every single part would be inspected.) That documentation was given to the NRC, where officials accepted it largely at face value.

At the 1979 licensing hearings, intervenors like Tom Carpenter and David Fankhauser were determined to show that this system was broken. They took time to explain the contentions of the whistleblowers to the Atomic Safety and Licensing Board and even convinced some of the whistleblowers to appear in person to read their affidavits. CG&E and NRC legal counsel hounded the men when they presented their statements, trying to show that the whistleblowers were either disgruntled ex-employees who had been fired or were unqualified to comment on Zimmer. At one hearing, after a whistleblower finished summarizing his concerns about the cable trays, CG&E's attorney, Troy Connor, challenged him, "You were responsible for the people who certified the welders. You let them do it?" The whistleblower replied that he had approved the certification, but he added, "I reported [my concerns] to the person I worked for." Connor pressed back, "When you were let go, it was then that you decided to report it." "Not true," the whistleblower insisted.[22]

At one hearing late in the year, CG&E brought in the director of the fire research department at Construction Technology Laboratories. This was CG&E's attempt to debunk whistleblowers' claims that Zimmer's cable trays were insufficient, either that they were being overloaded or were not durable enough. At Zimmer, the cable trays were being insulated with a product called Kaowool that had to meet a certain fire rating. If a fire broke out at the plant, CG&E needed to know that cable trays would be able to withstand it and support the vast system of electrical wiring.[23]

In off-site laboratory tests, Construction Technology had simulated a fire around Kaowool-insulated cable trays—similar to the installation at Zimmer—to make sure that temperatures tested along particular junctures of the cable trays did not exceed certain limits. As long as the average of the sampled temperatures did not surpass a set temperature, then that specific combination of Kaowool-insulated cable trays met the fire rating and could be used at Zimmer. That is what happened during tests. Construction Technology's fire research director said as much to the Atomic Safety and Licensing Board—that, given these successful tests, everything would be fine at Zimmer as long as workers installed insulation and cable trays as computer data and architectural designs told them. One CG&E official reminded everyone that if

Kaowool was consistently made and tested, "it's the results of the test that are significant"—not the product itself.[24]

Intervenors and whistleblowers argued such testing standards were too lax, opening the possibility for unsafe cable trays. Instead, these parties insisted, Kaowool and other products and parts should undergo redundant laboratory testing and multiple on-site inspections after installation. As David Fankhauser put it, "When it comes to nuclear power plants, we cannot permit safety to barely be satisfied." But this wasn't the way of the industry. For instance, with the control rods, CG&E planned to regularly check them after Zimmer was in operation for a certain period of time. Intervenors asked, what if the control rods acted abnormally prior to the start of the tests, as a result of "cracks that may be induced by somewhat improper construction practices?" CG&E counsel responded that that was not possible.[25]

At the 1979 hearings, CG&E also solicited appearances from engineers, contractors, construction workers, and inspectors at Zimmer—the colleagues of the whistleblowers—who insisted that everything was fine at the plant. Devine wasn't surprised to learn this. For CG&E, this was clearly a tactic to discredit the whistleblowers, but Devine understood that it was also a way for Zimmer workers to defend their professional reputations and their jobs. After all, if Zimmer was canceled or otherwise jeopardized, the high-paying work there would vanish. Whistleblowers knew that their fellow workers deeply resented them; some even shared that they had been threatened. At one 1979 hearing, a woman from the audience took the podium to share such a story. She and her husband drove to Zimmer earlier in the year to see the plant, and when parked outside of it, someone tapped on their window. "We opened the window a crack," she said, "and it turned out to be a guy who was a worker at Zimmer and who had, you know, come off shift. He wanted us to open the door and get into the car and talk to us. We were scared to do that. There were so many stories about hardhats being intent on preserving their jobs at all costs, those kinds of stereotypes." The couple decided to let the worker in their car. "And for about two hours," she described, "he told us about the conditions in the Zimmer plant, about the low morale there, about the drugs and about accidents which seemed to constantly occur and which often they didn't seem to be able to deal with immediately." He shared news of a leak involving "a number of terminals"—"where they didn't know how to turn off the water for a long time. And we asked him, why don't you give public testimony about what's going on in there?" He explained he was scared of losing his job. "He even mentioned the possibility of physical reprisal," she said.[26]

As Devine acquainted himself with this backstory of Zimmer, he understood that whistleblowers—Applegate and the others—were a minority at Zimmer (as whistleblowers usually are), but he also knew that across the US, some tradesmen building and operating nuclear power plants were becoming more critical of the technology. Workers were expressing concern over associated health risks. Some men even participated in antinuclear demonstrations. Coming short of outright opposing nuclear power (since that would have meant opposing their own jobs), they instead demanded exhaustive NRC regulation so that they would have safe working conditions. At one march in Washington, DC, in May 1979, the president of the International Association of Machinists told the crowd, "Workers slowly die of radiation making nuclear fuel and operating nuclear power plants. Workers transport the fuel. [They] are frontline economic, health, and safety casualties whenever nuclear accidents and mishaps occur." Relatedly, workers at nuclear weapons production sites were also worried about routine exposures to radiation. A 1978 study by Thomas Najarian, a hematology fellow at the Veterans Administration Hospital in Boston, had shown that nuclear submarine workers at the Portsmouth Naval Shipyard in Maine were nearly five times more likely to die from leukemia than the public. In response to Najarian's study and others, workers at various nuclear facilities began requesting radiation exposure surveys.[27]

Perhaps this activism buoyed the whistleblowers in Cincinnati. Regardless, they certainly could count on a supportive audience at the 1979 hearings. Members of the public there—shaken by the stories whistleblowers were sharing—spoke in support of men coming forward and shared their own news of negligence at the power plant. One woman told the Atomic Safety and Licensing Board, "I know several people who work in the Zimmer plant. One is a foreman who says that there are *Playboy* type centerfold pictures decorating the walls, that men sleep in the cable trays . . . and that workers carry coolers of beer to work instead of lunch boxes." Stan Hedeen, the conservation chief for the Ohio chapter of the Sierra Club, toured the plant after CG&E, trying to court a better public image, invited certain community leaders. Hedeen was struck by the large amount of graffiti throughout Zimmer. Afterward, he asked himself, "If this was their level of care of what they were building, then what is the quality of what they're building?" In the early 1970s, Hedeen attended Zimmer's construction permit hearings and there advised caution against the plant until its environmental consequences were better understood. By the late 1970s, he said he was not surprised when the whistleblowing started.[28]

Devine was encouraged by this popular support for whistleblowers—and the Cincinnati area's increasing unease over Zimmer. With his growing body of evidence that the utility and its general contractor were inadequately building and managing Zimmer, Devine dug deeper, going to the ultimate source for more information—the NRC itself.

After initiating his work on Zimmer in the summer of 1980, Devine spent the rest of the year and the first months of 1981 using the 1966 Freedom of Information Act (FOIA) and 1970s FOIA amendments to procure copies of all NRC inspections of Zimmer beginning in March 1979. He also secured transcripts of private NRC meetings and telephoned NRC officials. This information—along with statements from Applegate and other whistleblowers—painted an indicting picture against CG&E and the NRC's management of Zimmer.[29]

In May 1981, Devine sent a summary of his findings to James Keppler, the Chicago regional director of the NRC, who was overseeing Zimmer. In the summary, Devine concluded that according to the NRC's own rules, it should halt Zimmer's ongoing construction. In an exhaustive level of detail, Devine identified "62 items of noncompliance with law or NRC regulation over the last two years." He wrote that as of December 1980, when more than 90 percent of safety-related components had been installed, over 50 percent needed repairs: "In addition to repeated noncompliance and defective components in safety-related systems, NRC reports, combined with our own investigation, have documented mismanagement and systematic defects in two vital areas—1) the integrity of the quality assurance program, and 2) adequate security measures to protect the plant and the nuclear fuel already there." He told the NRC that the only appropriate course of action was to stop Zimmer's construction and restructure its quality assurance program. To do otherwise, Devine said, was a serious threat to public health and would only result in a larger mess in the future. (NRC rules specified that it could revoke or halt a license if public health, common defense or security, or the environment was at stake.)[30]

Through subpoenaed NRC documents, Devine made the case that the commission's *own records* showed that Zimmer's construction and quality assurance were profoundly compromised. The initial wave of whistleblowers who came forward in the mid- and late 1970s had prompted NRC investigations, and while the commission publicly insisted that everything was either fine or fixable at Zimmer, Devine demonstrated that the NRC's own documents contradicted this. One millwright, for instance, claimed in 1979 that small metal shavings were possibly clogging the control rod blades even though his company passed the blades in their tests. NRC inspections confirmed that the millwright was correct. Another whistleblower had alleged in 1979 that control rods were deficient and incorrectly tested. The NRC confirmed that 86 of 137 control rod blades failed to pass certain tests. Eleven exceeded permissible limits of thickness under pressure. Since control rods allowed operators to slow down or stop the fission in the reactor, they were vital parts. If the blades on them were larger than specifications allowed, the rods could fuse with the reactor vessel if high enough temperatures were observed during some kind of malfunction, possibly resulting in a meltdown. And the allegations of a quality assurance inspector—that the control rod drive pump, which activated the control rods, was missing at Zimmer—turned out to be true. NRC officials reported that it was eventually found in a pile of mud under a construction trailer on site. After locating it, workers did not test the pump to verify that it could run long term.[31]

Beyond these issues, Devine reminded the NRC of its record of violations at Zimmer—fifty-nine violations over construction and inspection practices from 1979 to 1980 alone. Issues ranged from the quality of the parts, to the quality of the work, to the quality of the workers. The NRC, for instance, had found that welders were unqualified and were, to no surprise, performing substandard work. One inspection revealed that workers had not protected already installed cables when they were welding nearby, which could have damaged the electrical cables to the point that they would short-circuit or start a fire. Devine pointed officials to their own words, explaining, "In short, the NRC reports evidence of a systematic breakdown of the welding process at Zimmer, including personnel training, equipment, records and inspections. There is little question that the confirmed defects only hint at the magnitude of flaws produced by this operation. And the defects in welds prepared outside the plant may dwarf the in-house defects." With regard to piping and hangers—"the skeleton of the system"—the NRC revealed the piping was deteriorated and that CG&E was noncompliant with safety standards.[32]

Devine confirmed—again, through the NRC's own records—shoddy maintenance of equipment, poor sanitation, and haphazard workmanship practices. "Unfortunately," he wrote, "NRC reports on the Zimmer construction site describe many of the characteristics of a garbage dump." Floors were dusty and dirty. Construction debris—goggles, concrete, wood, pipes, soda bottles, empty oil cans, apple cores, light bulbs, wire brushes—was littered around. The NRC ordered CG&E to clean repeatedly, but its records showed that each time its staff returned, Zimmer was still untidy. Devine made clear that aside from being a messy job site, the extraneous items and debris were hazards to ongoing safety-related construction work. He reminded the NRC that "the inspector found excessive amounts of combustible material in the Service Water Structure, and light bulbs in a cable tray. Most significant, a collection of beer cans and paper trash was observed . . . in [a tunnel] . . . of the reactor building." Citing another NRC report to the commission, Devine described workers at Zimmer: "When they broke for lunch, pipe fitters forgot to close the valves and stop the flushing that operates heat removal systems for the reactor. The consequences were frightening. Water overflowed from the reactor building equipment drain tank and into the corner room which contained the pumps, instruments and valves." The room quickly gathered five feet of water, which had to be drained.[33]

Devine also relied on his own investigative powers to verify the NRC's findings. This was easy since whistleblowers continued to come forward to him, identifying issues with design and construction work as well as the qualifications of tradesmen on the job site. Workers leaked to Devine that architectural plans were drawn *after* construction had commenced to comport with work already done, and even then, key items did not match design specifications. They also said that engineers did not approve certain design criteria, including key structural work items like the selection of steel beams. NRC records verified this. Other whistleblowers told Devine that there was a design flaw in the control panel of the heat exchanger. A heat exchanger transfers thermal energy from a reactor's boiled water to steam and, later, from steam back to cooled water. The design error at Zimmer resulted in 1,200 pounds of water pressure circulating through pipes designed to handle only 300 pounds. As a result, workers observed pipes rippling and finally bursting, water spraying electricians in a powerful stream. "My god—it was scary," the whistleblower said.[34]

Men shared with Devine serious issues with the welds at Zimmer, both the actual work and the qualifications of welders. One man told him that welding at certain vital junctures of cables was inadequate. Devine reported this to NRC, saying, "In the event of a weld failure at a juncture, the cables would fall and possibly have their insulation cut off due to sharp edges where the tray breaks. At a minimum there would be an electrical short. More likely, he [the whistleblower] explained, the result would be man-made lightning." One of Zimmer's earliest whistleblowers came forward again, telling Devine that welders were allowed to retake qualification tests as many times as needed until they passed. One man took the test sixty times before he was certified. An NRC record substantiated this, finding that the work of seventeen of thirty welders in the summer of 1980 had been rejected as poor quality. Devine argued in his correspondence with the NRC, "In our opinion, spot-checking is an inappropriate response when over half the welders checked in this sample could not pass muster." There was no documentation that welding equipment met standards of the American Society of Mechanical Engineers, and on certain inspections, the NRC identified serious defects in the equipment. Even worse, the NRC had already reprimanded CG&E numerous times for these various issues, yet as Devine learned, the utility was not improving in its management. The NRC wrote in the fall of 1980, "The corrective action taken to ensure control of weld rods used for welding of safety-related components has been inadequate."[35]

Devine pointed out that construction errors would exacerbate the issue of human error in Zimmer's operations. Even if everything was installed perfectly at the plant, people still operated it, opening up a window for an occasional blunder. Given that inadequately trained workers had built major components of the plant, human error at Zimmer—"the Achilles Heel of even the most well-constructed plants," as Devine wrote—was more likely. Many workers were, as Devine noted, "transient workers" who jumped around temporary jobs at nuclear power plants. Beyond that, alcohol and drug use was common on the site. Whistleblowers shared stories of red-eyed employees "who did their work under the handicap of an alcoholic stupor." They said cocaine and marijuana were also commonly used on the job and that when security guards brought these issues to CG&E leadership, the men were told to throw away the drugs so that CG&E officials did not have to file any paperwork over the matter. "A nuclear plant constructed by drunken employees is more likely to stagger into an accident," Devine wrote. Beyond all this, there were also workers on parole for serious drug possession and even some for murder, and it was known that CG&E and Kaiser did not conduct background checks prior to hire. One whistleblower told Devine that a fellow worker carried around a machine gun and ammunition in the back of his car.[36]

The security situation was also compromised. Due to repeated bomb threats at nuclear power plants across the US—including some at Zimmer—this was an urgent issue. From workers' reports and meetings Devine held with NRC officials, the GAP attorney concluded that at Zimmer, "employees have described a system that is little better than a joke, with drunken, overworked, untrained and poorly supervised, understaffed and poorly armed guards, the strongest defense the plant has against attempted sabotage. According to one former guard, the security system represents a 'showpiece' but nothing else." The security force guarding Zimmer's stored nuclear fuel—two-men strong—often worked for forty-eight hours straight, which meant they often had to leave the fuel unattended to take a break. Their supervisor was perpetually intoxicated. They had no weapons training and passed firearm tests only by retaking them multiple times. Elsewhere on site, men routinely broke into secure areas with keys and credit cards.[37]

Devine exposed additional criminal activity at Zimmer. Applegate had first raised these issues, and Devine vindicated him with the affidavits of several other employees at Zimmer. Workers sold stolen guns on site. One superintendent had stolen and resold materials from Zimmer for his personal benefit, costing CG&E more than thirty thousand dollars. Others had made belt buckles from nuclear-grade steel and sold them for personal profit. Thirty employees jointly smuggled two thousand pounds of copper cable out of Zimmer and sold it for fifteen thousand dollars on the black market. They used that money to finance an extravagant Christmas party, which included hiring prostitutes. Devine argued that these crimes constituted legitimate cause for the US Department of Justice to get involved.[38]

Finally, Devine highlighted the issue of missing documentation. CG&E's quality assurance records were insufficient, inaccurate—and often, just not there. Required independent audits did not comply with NRC standards or had not even occurred. Most problematic was that CG&E and Kaiser were supposed to document and report issues of noncompliance immediately to the NRC, but as Devine figured out, the utility and its contractor had instead tried to cover them up. "The utility and Kaiser routinely have fired employees for 'incompetence' who persisted in voicing safety concerns," Devine explained to the NRC. Even GAP was the subject of a CG&E attempt to discredit opposition. With the Cold War ongoing, the utility accused the law firm of having communist sympathies: "CG&E officials have even stooped to a baseless, defamatory campaign of 'red-baiting' the Government Accountability Project in retaliation for GAP's assistance to Mr. Applegate." "In short," Devine summarized, "there is serious question whether the Zimmer plant 'on paper' is the same as the Zimmer plant in reality."[39]

In contrast to CG&E's quality assurance documentation, the paper trail that Devine created was exhaustive, and it condemned Zimmer. With painstaking detail, Devine identified all the safety issues, analyzed them, and summarized them in an orderly fashion for the NRC to see how dangerous Zimmer was. He was careful, though, to not critique nuclear technology at large. He argued that as long as CG&E left the issues at Zimmer unresolved, "the citizens will be further delayed in enjoying the benefits of nuclear power from Zimmer." Devine was cognizant of the fact that he worked for a nonpartisan law firm and that speaking against nuclear energy on principle would not help his case with the NRC. Furthermore, many of the whistleblowers he worked with remained "pro-nuclear," as Devine later said. Nuclear power, after all, gave them jobs. But they told Devine that it was a technology that required intensive management to make it safe, and they did not feel that CG&E, its general contractor, or the NRC was providing that assurance. Devine painstakingly built his case against Zimmer, and Zimmer alone—a careful choice that enabled him to challenge CG&E and the NRC while still keeping the confidence of whistleblowers.[40]

While preparing his report to the NRC, Devine pursued another angle, too, to make sure his work wouldn't be dismissed. As part of the Civil Service Reform Act of 1978, government workers could blow the whistle formally through the US Office of the Special Counsel, an independent federal investigative and prosecutorial agency. Outside of government workers, if the counsel received information that supported a reasonable belief of a threat to public health and safety, it had the power to refer that information to the appropriate federal agency for investigation. By the time Devine became involved with Zimmer, no private citizen had ever used the whistleblowing provisions in the Civil Service Reform Act in that way. Still, he gave it a try, informing the Special Counsel that CG&E's mismanagement at Zimmer and the NRC's avoidance of the issues constituted a violation of law, an abuse of authority, and a danger to public health. It worked. The Special Counsel referred Devine's evidence to NRC chairman John Ahearne, who reopened his commission's review of Applegate's original claims. Altogether, these developments prompted two investigations by the NRC: one by its Office of Inspector and Auditor and another by the regional team in Chicago (which, again, was in charge of Zimmer). Bureaucratic rivalries helped to propel these inquiries. The Office of Inspector and Auditor—then trying to flex its power over the Chicago office—went to work. Worried about being under the microscope, Chicago conducted its own probe into Zimmer, interviewing Applegate and other whistleblowers. These investigations lasted throughout 1981.[41]

Finally, on Tuesday, November 17, 1981, in response to Devine's work and the dual NRC reviews on Zimmer, the NRC's Office of Inspector and Auditor released a report in which it confirmed that the Chicago team had failed to regulate Zimmer properly and thoroughly investigate Applegate's allegations. Its report stated, "Applegate was correct in saying that defective welds in safety-related systems had been accepted. . . . The overall finding is that the [first NRC] investigation . . . was unsatisfactory." It was the first time that the NRC publicly repudiated its own at Zimmer. Applegate commented at the time, "After they called me a nut for over a year, they've finally confirmed I was right all along." Devine summarized the investigation by saying that "the NRC confirmed what critics had been charging for over a decade: NRC investigations are little more than paperwork reviews."[42]

Then, on Wednesday, November 25, 1981, about a week later, the NRC's Chicago branch finished its investigation by levying a $200,000 fine on CG&E for a widespread breakdown in its quality assurance program. It was then the largest fine in the NRC's history against a nuclear plant under construction. Suddenly, Cincinnati was on the *ABC Nightly News*.[43]

In response to the fine, the director of the Chicago office, James Keppler, held a press conference in downtown Cincinnati's Netherland Plaza Hilton Hotel, not far from Stouffer's Hotel, where CG&E executives had so proudly announced their Zimmer project in 1969. Twelve years later, the mood was quite different. NRC officials arrived at the press conference armed with a two-inch-thick report detailing 950 examples of noncompliance and 1,000 violations at Zimmer. Showing falsified quality assurance records, physical harassment of inspectors, and nonconformance with safety requirements, the report was damning. It showed to the public, in no uncertain terms, that even if the NRC did not substantiate the claims of each and every whistleblower, the vast majority of allegations were corroborated, justifying and validating the existence of whistleblowers. As such, Keppler explained that the NRC was mandating a total plant "recheck" by both his commission and CG&E, given that the project was—in his words—"totally out of control." While the NRC stopped short of halting construction, it did maintain that even though Zimmer was so close to completion, no fuel could be loaded. Furthermore, Keppler made clear that if CG&E's quality assurance program failed a second time, the NRC would consider suspending the construction license. The NRC gave CG&E until February 1982 to respond to the charges and the fine.[44]

At the end of the day, when reporter Ben Kaufman filed his story for the *Cincinnati Enquirer*, he called Zimmer a "dance of the virgins," in reference to the inexperience and mismanagement of all parties involved. CG&E president William Dickhoner disagreed. To Kaufman, he said the fine was excessive, unjustified, and unnecessary and that his company was being subject to an "unusually high degree of scrutiny." After all, in his mind, CG&E was merely being fined for a lack of records. "It's really a situation where we have been fined for documentation," he stated. He was correct only in the sense that his company had profoundly failed to maintain a proper paper trail for Zimmer's quality assurance. As to the criminal activity that Devine had found, Dickhoner and his colleagues denied this, saying, "We took great issue with that." And the rest of the charges by Devine and the NRC? They were, in Dickhoner's words, "nothing of significance."[45]

"Zimmer was not an aberration," Tom Devine said about forty years after the NRC slapped its fine on CG&E. "It was representative of nuclear safety regulation that was for appearances and as a liability buffer rather than as a public health and safety law enforcement priority."[46]

In 1979, the NRC received reports for over 2,300 mechanical failures and human errors at licensed nuclear power plants, and the following year, when there were almost 70 licensed nuclear power stations across the US, the number of incidents was 3,804.

Examples of errors included control rods not working, an electrical system failure that left operators in the dark, a coolant pump seal blowing out, a steam bubble forming in the reactor system, and loss of coolant in a reactor. Equipment failures caused the majority of problems, but human error amounted to 20 percent—and these were just the reported incidences. As Zimmer showed, many utilities did not report accidents and mishaps to the NRC. To this point, NRC officials were pleased that the number of incidents in 1980 had risen from 1979 since it showed that the regulatory process, in their words, "was working."[47]

In 1981, the NRC acknowledged that in more than a dozen nuclear plants, exposure to radiation was making the steel shells that surrounded reactor cores so brittle that some of the plants might have to be shut down for repairs. That year, Oak Ridge National Laboratory in Tennessee—a Department of Energy–owned and contractor-operated facility developing reactors—released a study for the NRC, citing 19,400 incidents and failures in nuclear power plants from 1969 to 1979. The NRC commissioned the study after it endured backlash for its 1975 Rasmussen Report, which, focusing on one reactor in Virginia, concluded that multiple equipment failures would have to occur before an accident could damage the core and that the chance of such an accident was one in twenty thousand years of operation. Critics questioned such an optimistic projection, and trying to legitimize the safety of nuclear power, the NRC turned to Oak Ridge for an independent, and hopefully similar, report. But the Tennessee lab didn't conclude the same: instead, its study estimated the likelihood of a major accident at a nuclear power plant to be one in one thousand years of reactor operation. Beyond that, the report concluded that of the 19,400 malfunctions in nuclear power stations from 1969 to 1979, 169 of them had potentially contributed to serious accidents. The Oak Ridge authors said that given the state of equipment, it was surprising that more serious accidents had not occurred. They concluded that for those years, a major incident could have been expected every ten to fifteen years. And although they said that in the future, America's reactors could expect a massive problem every one thousand years, the Oak Ridge report was nonetheless "rushed to [NRC] commission staff personnel and others involved in special hearings in White Plains, N.Y., that are considering whether the Indian Point [nuclear] plants should be closed for safety reasons," as Matthew L. Wald of the *New York Times* reported. Indian Point, just north of New York City, along the Hudson River, was one of the oldest nuclear power plants in the US, and as early as 1974, the NRC had shut down one of its reactors due to a design defect in its emergency cooling system.[48]

Why were all of these issues happening? To begin with, utility executives managed the construction process for nuclear power stations largely the same as their other power plants. They failed to appreciate that, compared to fossil fuels, nuclear power required extensive supervision and quality assurance as part of its construction management process. Power companies and the general contractors they hired, like Kaiser at Zimmer, were also working with tradesmen who had built fossil fuel–powered plants, not nuclear power plants. Workers—and their higher-ups—struggled to understand the importance of the extensive documentation and inspection process needed to ensure public and worker safety. As one activist who organized against Zimmer's licensure later recalled, he had the "gnawing sense that CG&E was out of its league." Additionally, the tenuous economy of the 1970s and early 1980s drove utilities, desperate to see a return on their nuclear power plants, to prioritize speed and cost savings over public health and workers' safety. Furthermore, the Atomic Energy Commission and the Nuclear Regulatory Commission's way of devolving crucial components of regulation made these issues of utility inexperience and negligence even worse. On top of that, faulty technology also plagued nuclear power at this time. During the bandwagon phase of early commercial nuclear power in the 1960s, reactor design was new and imperfect, and yet by the following decade, General Electric, Westinghouse, and other reactor manufacturers continued to identify a number of safety issues with their models, requiring corrections. Through all this, power companies maintained their faith in the technology and in their managerial experience. This made them very reluctant to acknowledge any issues with nuclear power.[49]

And so it was whistleblowers who alerted the NRC and the public to these problems. In the summer of 1979, Charles Cutshall, a concrete worker at the Marble Hill nuclear plant in Madison, Indiana, submitted an affidavit to a local group organizing against the plant. In it, he reported incorrect concrete mixing for the reactor's containment building. Then other workers there came forward with safety-related construction issues, prompting NRC review. James Keppler of the NRC confirmed Cutshall's charges: "Everything Cutshall has said so far has turned out to be true and we don't take his comments lightly. The NRC investigation will now center on how much Public Service Indiana [the utility] knew about the defects, and what it did with that information." The NRC stopped safety-related construction work as it investigated the issue. By October 1979, officials announced that Public Service Indiana was responsible for poor construction at Marble Hill, citing "more problems at Madison than anywhere else." As local protest mounted against the plant and even included state representatives, the FBI worked with a House subcommittee on an "investigation of alleged conspiracy and cover-up of the construction defects" at Marble Hill.[50]

Around the same time, a quality assurance inspector employed by the construction company Bechtel came forward to the NRC to say that welding work at both the Midland, Michigan, nuclear power plant and the one in San Onofre, California, involved thousands of substandard welds on pipe and electrical system supports. The Bechtel worker said it was probable that a significant accident could occur as a result. His whistleblowing came after forty years of experience in welding, seventeen of which were also in quality assurance. Bechtel was then the world's largest private construction company and the builder of many nuclear power plants. In response to the whistleblower's statements, the company fired him.[51]

At the same time, workers at the Diablo Canyon plant in California identified serious construction mistakes there. This came on top of the issue that the plant was being built on the Hosgri earthquake fault, as confirmed by Shell Oil and the US Geological Survey. Despite these issues, the NRC approved low-power testing in 1981, resulting in mass protests near the plant.[52]

Like with Zimmer, many whistleblowers turned to legal counsel, who helped them go public. After workers at the Seabrook nuclear plant in New Hampshire came forward with serious safety-related construction issues, they turned to the Employee's Legal Project for support. The Government Accountability Project itself represented whistleblowers at the Midland and Diablo Canyon plants as well as the nuclear station near Ottawa, Illinois. Tom Devine also represented a nuclear engineer at Three Mile Island who had safety concerns about Bechtel's remediation of the reactor core following the accident (namely, that it was using a structurally compromised crane to remove the reactor).[53]

Looking back on the 1970s, whistleblowers stand out as a key fixture. The extent of the failed Vietnam War, the Watergate scandal—both "leaks" in the early 1970s were because of whistleblowers (and, of course, sharp investigative reporting). The very creation of Tom Devine's law firm, along with other public interest ones, suggests that whistleblowers were numerous enough by the 1970s to devote an entire legal practice to their protection. And while whistleblowers weren't unique to nuclear technology, nuclear power plants did produce many. So did nuclear weapons production facilities, where in the late 1970s and through the 1980s employees accused the Department of Energy and its contractors of not prioritizing workers' and residents' safety. The most famous whistleblower to emerge from nuclear weapons production in the 1970s was Karen Silkwood. She was an employee and union organizer at the Kerr-McGee plutonium-processing plant in Oklahoma. After finding plutonium on her body and in her home, she raised safety concerns about workplace practices. In late 1974, on her way to share this with a *New York Times* journalist, she died in a mysterious car accident that some concluded was a calculated murder. The story received widespread coverage and was made into a 1983 film, in which Meryl Streep played Silkwood. (When Thomas Applegate and Tom Carpenter were driving to Washington, DC, to meet with Tom Devine for the first time, Applegate brought a handgun with him because, in his words, he was afraid something was going to happen to them "Karen Silkwood-style.")[54]

The emergence of whistleblowers in the 1970s played into the impression, held by a growing number of Americans, that federal agencies and corporations weren't being transparent or acting in the public's best interest. To the communities around contested nuclear power plants, where whistleblowers were coming forward, NRC officials and utility managers seemed to be perfect examples of this accountability problem. When, in the summer of 1980, the NRC reported over 2,300 mechanical failures and human errors at licensed nuclear power plants, the head official analyzing these reports said, "I would attach no significance whatsoever to [the number of reports]."[55]

This wasn't the response people near nuclear power plants were looking for. Instead, it was clear in places like Cincinnati that the public wanted the NRC to step up. After its $200,000 fine and mandated recheck of Zimmer, Ben Kaufman wrote in the *Cincinnati Enquirer*, "It's what critics have been demanding for years. They never accepted the NRC's explanation that federal inspectors function mostly as auditors, checking a sample of work while relying on quality assurance documents provided by CG&E and its contractors."[56]

People in Cincinnati would have to wait and see what exactly the NRC would do with the recheck, if officials would stop acting like mere auditors. In the meantime, CG&E began to host tours of Zimmer for local journalists, believing that they were spreading misinformation and exaggerating how dangerous the errors at Zimmer actually were. The lead superintendent there explained that "even minor incidents get completely blown out of proportion, simply because [news reporters] do not understand." But *Cincinnati Post* and *Enquirer* reporters, those on the environmental beat, were not inventing facts about Zimmer or taking the side of the Zimmer opposition. They simply reported. And by the time of the $200,000 fine, many were exasperated with CG&E public relations staff. Kaufman later said, "It got so bad at Zimmer that reporters who followed the fiasco began calling CG&E PR people 'designated liars.' I don't know if they were lying or mouthing what they were told [by higher-ups]."[57]

Altogether, the ways that utility companies and their contractors prioritized speed and costs over safety—and the ways that the NRC was reluctant to change its regulatory style—encouraged people who lived near Zimmer and other contested nuclear power plants to further mistrust the authorities in charge of them. This, it turned out, proved to be a great motivating force, contributing to growing local movements against what were perceived as unsafe power plants. Indeed, as whistleblowers raised *specific* safety concerns over *specific* nuclear power plants, more people nearby them took notice. The issues, after all, suddenly seemed pertinent and urgent to those communities in a way they hadn't in years prior. Nationally, the effect of this was that by the early 1980s, Americans opposed the construction of a local nuclear power plant by a 2–1 margin, a complete reversal from the decade before.[58]

On top of safety concerns, cost issues also loomed large in communities home to nuclear power plants. By 1980, Zimmer had already eaten up $1 billion of CG&E and its partners' investment and was years behind schedule. These issues might have not mattered that much to most CG&E ratepayers, but in that same year, 1980, CG&E and its co-owners got permission to include Zimmer's costs in the rate base they were charging customers—an outcome that started occurring at other nuclear power plants across the country. Suddenly, with their monthly bills now reflecting the costs of a faltering power plant already 400 percent over budget, people started to take notice.[59]

6

"It's Time to Stop Bailing the Utilities Out"

Ratepayers against Zimmer

On Friday, March 28, 1980, exactly one year after the Three Mile Island accident, Cincinnatians woke up to a page-length spread in the *Cincinnati Enquirer*: "The Zimmer Nuclear Plant: A Risk Cincinnati Can't Afford." The article stated in no uncertain terms that "the people of Cincinnati will be better off if Zimmer never opens." Tom Carpenter's Cincinnati Alliance for Responsible Energy and the Zimmer Area Citizens groups, who had written the opinion piece, wanted to warn readers that CG&E would continue to request and receive rate increases from the Public Utilities Commission of Ohio to pay for Zimmer. "Zimmer = rate hike + rate hike + rate hike," they wrote.[1]

This was a new issue. While CG&E had procured rate increases throughout the 1970s, it was not until 1980 that Zimmer was explicitly part of these. Early that year, the Public Utilities Commission authorized construction costs from the nuclear power plant to be included in CG&E's electric rate base. Even though the plant was not yet operational, there was a state provision known as construction-work-in-progress, where if a power plant was at least 75 percent complete, the PUCO could at its discretion include some of the plant's construction costs in the base rate. (A caveat was that construction-work-in-progress could not raise customers' rates by more than 20 percent of their original rate.) Commissioners also allowed the same for Columbus and Southern Ohio Electric and Dayton Power and Light, so across central and southwest Ohio—large portions of the state—ratepayers were suddenly paying for Zimmer.[2]

At the bottom of the *Enquirer* op-ed, around three hundred people listed their names in support. Some were unsurprising, old foes of Zimmer and CG&E at this point. These included city of Cincinnati council members, whistleblowers and other former employees at Zimmer, and members of CARE and Zimmer Area Citizens. But many names were new to the cause, showing how "Zimmer = rate hike + rate hike + rate hike" was galvanizing a new section of regional residents. CG&E, Columbus and Southern, and Dayton Power customers put down their names. So did church leaders, teachers and university academics, attorneys, businesses, neighborhood councils, local chapters of national peace and social justice organizations, and major civil rights leaders like Reverend Fred Shuttlesworth. The fact that ratepayers had to pay for Zimmer—a power plant that was struggling to gain licensure, that was years behind schedule and two times over its original budget—struck many people as, frankly, unfair. And this realization, like Three Mile Island and the whistleblowers' affidavits, mobilized people in a new and urgent way.

Stemming from earlier anti-CG&E organizing, new groups formed to protest rate hikes and Zimmer's inclusion in them. There was the Citywide Coalition for Utility Reform, made up of working-class and low-income ratepayers. They argued affordable energy was a racial and economic justice issue, much like Jesse Jackson's People United to Save Humanity did when it organized against CG&E's billing practices following the 1973 energy crisis. There was also the Ohio Office of Consumers' Counsel. Formed in 1976 by public interest lawyers, the counsel intervened at Public Utilities Commission hearings on behalf of Ohio utility customers, fighting against rate increases. And then there was the Ohio Public

Interest Campaign, which attacked the construction-work-in-progress provision and stirred up more agitation among Cincinnatians against CG&E's expensive nuclear power plant. None of these groups focused on Zimmer's safety-related issues. Instead, by concentrating on fair utility practices and Zimmer's financial impact on customers, these groups engaged previously unengaged residents, growing another arm to the anti-Zimmer movement. Against this local backdrop was a burgeoning national movement for consumer rights, and across many states, including Ohio, calls for utility reform emerged part and parcel with it.

At the same time, intervenors in Zimmer's licensing hearings—people like David Fankhauser—also targeted Cincinnati Gas and Electric for Zimmer's high costs and, to them, CG&E's precarious finances. When the first licensing hearing since 1979 was convened in the spring of 1981, a few months before the Nuclear Regulatory Commission leveled CG&E with the $200,000 fine, an Atomic Safety and Licensing Board listened as intervening parties took CG&E and its partner utilities to task for not considering how they would handle unexpected and financially deleterious situations (i.e., if there was an accident at Zimmer, would CG&E and its co-owners be able to cover their portion of the costs?). On top of that, there was another recession, high inflation, declining bond ratings, reduced customer demand, growing interest in energy conservation, and a slew of nuclear power plant cancellations across the US, all of which signaled that utilities were no longer the powerful industry they once were and that they needed to adjust their business model. Yet CG&E officials remained confident and stubborn that Zimmer was still a financially prudent move. When the NRC agreed with CG&E, intervenors pointed out that Zimmer's cost issues, just like its safety issues, were widening the gap between the public and the federal commission that was supposed to be protecting them. The outcome of this, along with the anti-CG&E organizing by ratepayers, convinced more people to oppose Zimmer.

At the start of 1980, the Public Utilities Commission of Ohio granted CG&E an electric rate hike, which meant an increase of $35.1 million in revenue for the company. Five months later, Cincinnati Gas and Electric filed for another electric rate hike to add an additional $55 million of revenue to the company's finances. As the utilities commission considered the request, CG&E presented another in January 1981, this time for a $100 million increase in electric revenue and a $20 million increase in gas revenue. That March, the commission granted the company a $50.5 million electric revenue increase and another $12 million for unrecovered gas costs. Then in June, the PUCO allowed another $5.2 million revenue increase, raising electric rates further. When November came around and CG&E received its $200,000 NRC penalty, company officials filed for another rate increase.[3]

As a reminder, Ohio's Public Utilities Commission determined CG&E's rate base—and increases to it—by considering the company's total investment into power production (all their past and present incurred costs) and how much energy CG&E expected people to use, using consumption data from past years. In addition to these criteria, the PUCO awarded increases so that CG&E could maintain a solid credit rating, compensate stockholders, and collect a fair return on all it had put into producing power. (The Federal Power Commission, and later the Federal Energy Regulatory Commission, had the final word after PUCO granted, denied, or modified a utility's rate base.)

CG&E's repeated appeals for rate increases throughout the 1970s were a pattern that persisted, as we can see, into the 1980s. Cincinnati Gas and Electric executives argued that the slow approval process rendered frequent rate hikes necessary—that by the time rate increases were finally enacted (usually after a year or two delay), they were based on outdated incurred costs. There was also the poor economy, which, to company officials, further justified the recurring rate hikes. Indeed, the economic issues that arose with the 1973 recession didn't go away. In 1979, another oil shortage prompted a worldwide recession that lasted into the early 1980s, causing inflation to reach double digits.

The construction-work-in-progress provision, then, was a lifesaver to CG&E officials. In January 1980, the utilities commission permitted 50 percent of CG&E's investment in Zimmer in the rate base. CG&E executives saw nothing unreasonable about this. While leaders understood that customers were paying for an unfinished plant, they believed that once Zimmer was operational, consumers would greatly benefit from it. In fact, they told stockholders that Zimmer would save ratepayers more than $400 million over the course of its life. Furthermore, CG&E and its partners in Zimmer hadn't seen a penny back on their investment. It made sense to them that while they were expending significant resources, customers, set to benefit in the future, should share the burden in the meantime.[4]

CG&E ratepayers disagreed. And so they and their allies organized, building off anti-CG&E anger from the 1973 energy crisis. In 1977, the Citywide Coalition for Utility Reform formed from several block clubs across five Cincinnati neighborhoods (Over-the-Rhine, Mohawk, North Fairmount, East Walnut Hills, and Northside), all of which are close to the city center. In the 1970s, Over-the-Rhine, Mohawk, and North Fairmount were low-income neighborhoods with Black and Appalachian residents, while Northside and East Walnut Hills included a mixture of white and Black families from working-class and middle-class backgrounds. Drawing from these neighborhoods, the coalition had around eight hundred members and could consistently count on a few hundred people showing up to public meetings. Their mission? To improve CG&E's customer service and stop rising utility bills.[5]

"What we have in common," the president of the coalition, LaVerne Willsey, said of the Citywide Coalition for Utility Reform, "is that we are all poor." When CG&E received its electric rate increase in the summer of 1981, average customers saw their monthly bills rise by 13 percent. Residential customers who used 750 kilowatt-hours of electricity per month saw their bills go from

$33.38 to $37.78. For some, this was not much of an increase, but to families in Over-the-Rhine, Mohawk, North Fairmount, East Walnut Hills, and Northside, that increase *did* matter. Willsey said as much: "People are unable, literally, to meet their utility payments. It's a necessary service. It's necessary to life. Due to the fact that CG&E is a monopoly, you can't comparison shop." And because Zimmer's costs were included in the electric rate base and "because customers are paying for something they aren't getting," Willsey said her coalition opposed the nuclear power plant. It was the first time that significant numbers of Black and white working-class and low-income Cincinnatians organized against Zimmer, albeit not for safety reasons. To them, whistleblowers' concerns were just too remote to worry about given that they could barely afford to pay their utility bills. So instead, they organized under the banner of affordable energy, making it a matter of economic justice—and racial justice, too, given how many families of lower incomes were of color.[6]

Willsey was a Northside homemaker, then in her fifties. The middle child of three, she had grown up in Cincinnati; her father, George, worked for the city's recreation department, while her mother raised the family. Willsey went to Woodward High School in Over-the-Rhine; her junior year photo shows a young woman with wide-set eyes, light-colored and coifed hair, and an infectious smile. She later worked as a stenographer for a local grocery store before marrying. By the 1970s, she was a leading activist in her block club in Northside. There, she compelled the city to enact better police protection, litter control, and road enhancements for the area. After she saw her utility bill double from 1976 to 1977 and again in 1978, Willsey—no stranger to activism—organized her neighbors into a group that would fight CG&E on its rate increases. "Our position is that they [CG&E] don't need any more (money) than what they get in the base rate," she told the *Cincinnati Post.*[7]

In early 1977, Willsey and other coalition members attended public hearings convened by the PUCO in Columbus, Ohio, over

CG&E's fuel cost increases (which were included in its rate base). Willsey and her team petitioned to have legal standing in the hearings, arguing that they had a "specific interest" in the matter—that their utility bills were being affected by PUCO decisions. They were successful. Willsey also lobbied that daytime PUCO hearings regarding CG&E be moved from Columbus, about a two-hour drive from Cincinnati, to Cincinnati, giving CG&E ratepayers an easier way to air grievances. The PUCO's legal examiner denied the request, but on appeal to the commissioners, the coalition won. In the role of intervenor, and by moving the meetings to CG&E's hometown, the coalition challenged the rate increases and built a platform around which angry customers could gather and demand better treatment by CG&E.[8]

When the Public Utilities Commission traveled to Cincinnati in August 1978 to hold hearings in city hall chambers, the Citywide Coalition assembled hundreds of people there. At subsequent hearings in 1979 and 1980, the same thing happened. Holding anti-CG&E signs and wearing buttons that read "Freeze Profits, Not People" (in reference to the brutally cold winter of 1977), people pled to CG&E to pause power cutoffs during wintertime, enact lower late payment fees, and give people thirty days to pay bills. Some, even angrier, threw Monopoly paper money at Cincinnati Gas and Electric representatives. "I sit at night with the lights off watching TV," one older woman told the PUCO and CG&E officials. "I have a blood clot. I have diabetes. When the money won't stretch, I have to cut my medicines or my groceries." Another, a mother of ten children, said she could barely afford her utility bill, and another woman described her most recent electricity bill of $135 as "very depressing." Willsey told CG&E attorneys that their company's acronym did not mean "Cincinnati Gas and Electric" but rather "Customers give everything."[9]

Struck by the level of vitriol for CG&E, PUCO commissioners decided that the utility needed to hold its own public sessions in Cincinnati. It would be, they said, a bridge-building exercise in customer relations. CG&E representatives hesitantly agreed, convening a public meeting in December 1978 at a local Baptist church. Spearheaded by Willsey's coalition, around one hundred people came. Utility representatives heard various charges—"You have not read my meter in ten years"; "I gave you my whole paycheck and have nothing left over for anything else"; "I pay $125 more now that I've got storm doors and windows. And I'm still $500 in arrears and all I've got is a plain TV and a cooking stove." But as much as the coalition targeted CG&E, members were equally critical of the PUCO. Indeed, its reputation for being friendly—cozy—with utilities, more than a watchdog for ratepaying customers, endured. People, especially at public hearings, liked to call it "the puke-o."[10]

Cincinnati's city council, still in the hands of the Charterite-Democratic majority, was sympathetic to the plight of CG&E customers. As such, it connected Willsey's group with community development funds to aid its organizing efforts. Councilors also continued to contest CG&E's rate increase requests for city of Cincinnati customers and sent city solicitors to the PUCO's hearings, where, as intervenors, they fought Zimmer's costs from being included in CG&E's rate base. There were just too many outstanding issues with the power plant, they said, to permit customers to pay for it in the meantime.[11]

At the PUCO hearings, city solicitors were joined by lawyers from a new state agency: the Office of Consumers' Counsel. Created in 1976 by the Ohio General Assembly, the office served (and still serves) as a public interest advocate for utility customers. Its members were chosen by a nine-member bipartisan board, appointed by the Ohio Attorney General. Among other duties, they represented Ohio's ratepayers in front of the PUCO, federal regulatory agencies, appellate courts, and the Ohio state legislature. William A. Spratley was instrumental in the creation of the Consumers' Counsel. Spratley grew up in the 1950s in the northeast Ohio town of Wooster, went to the College of Wooster, and then attended Ohio State University for his JD. He became an attorney in 1973 and soon made a name for himself in public interest, environmental, and consumer rights law, to the extent that in 1977, he became the first director of the Office of Consumers' Counsel (a role he would hold until 1993). A prominent utility and energy reform advocate, he also served on a Department of Energy advisory board under President Carter in the late 1970s.[12]

Under Spratley's helm, the Office of Consumers' Counsel relentlessly fought the PUCO for fair rates for Ohio's utility ratepayers, winning a major victory in 1980 when Ohio's fuel adjustment clause was revised. The Ohio General Assembly had investigated this clause in 1975, finding that the provision (that utilities could include minor increases in the rate base without much fanfare) was defensible. The assembly did recommend that the PUCO better monitor its use across Ohio. The PUCO did so thereafter, instituting semiannual public hearings and an annual audit to determine if fuel procurement practices and adjustment changes were fair. But Spratley's team wasn't satisfied. In 1980, his office successfully petitioned the PUCO to begin mandating that a utility had to give six months' notice and justification to the PUCO prior to relaying *any* fuel cost adjustments to customers. It was a major victory to curb rising utility bills.[13]

In 1980, another public interest organization joined the fight for utility reform in Ohio: the Ohio Public Interest Campaign, usually called OPIC. OPIC was founded in Cleveland by a man named Ira Arlook. Born in New York City in 1943, the grandchild of Russian Jewish immigrants, Arlook went to Tufts University and then Stanford for a graduate degree in history. There, in the mid-1960s, he became a prominent organizer against the Vietnam War. After graduating, he worked with New England Resistance, an antiwar and antidraft group. Then, in 1975, he joined the faculty at Cleveland State University as an assistant professor. That same year, he started OPIC, its mission being to protect public health, improve environmental quality, benefit consumers, and promote democracy. He designed it as a mobilizing and organizing team, where campaigns started local and worked up to state and federal change.

Step one: OPIC staff went to a local community to motivate residents around public interest issues that particularly affected them. Step two: with local momentum, OPIC then built statewide campaigns with the power and wherewithal to lobby for legislative change at the state and federal levels. Its first campaign, spearheaded by Arlook, was Ohio legislation for manufacturing plant closures. Deindustrialization, outsourcing, and automation hit Ohio hard, especially the northern Rust Belt sections around Cleveland, and caused many manufacturing facilities to lay off workers with little notice. With the help of labor unions like the AFL-CIO and the Labor Council, OPIC successfully created and lobbied for a bill that required employers to give advance notice of plant closures to their workers. From that success, OPIC expanded its presence to Toledo and Columbus, working on different initiatives. Then it went to Cincinnati.[14]

Arlook wanted to create a grassroots movement against the construction-work-in-progress provision, and since it was particularly affecting CG&E customers, he thought Cincinnati was the perfect place to launch a campaign against it. But because the provision affected other utility ratepayers across the state, Arlook believed the campaign might result in a state moratorium. For this, he was hoping to court the support of Columbus and Southern and Dayton Power customers in Columbus and Dayton who were also paying for Zimmer. And he was banking on the support of Toledo Edison ratepayers in northern Ohio, who were paying for the Davis-Besse nuclear power plant—a plant with its own safety incidents and issues. To manage the organizing against the provision, Arlook recruited a person with a long history in activism and lobbying: Brewster Rhoads.[15]

Rhoads not only had the experience; he had the right temperament too. Known for his outgoing and affable nature and quick smile, he liked people, liked organizing them, and was good at it. Like Arlook, he became an adult at the peak of the Vietnam War. He was born and raised in Philadelphia, graduated from a Quaker high school in 1969, and then attended Williams College, a liberal arts college in western Massachusetts. There, he became deeply involved in civil rights, free speech, and the antiwar movement—to the point that he became the New England regional representative for the Coalition to Stop Funding the War, a major antiwar lobbying effort by peace, religious, and labor groups. He also worked as an antipoverty activist in the federal VISTA program—basically, the domestic version of the Peace Corps—helping to organize a local drop-in center near his college. After graduation, he moved to Washington, DC, in 1974 to work for, and eventually direct, the Coalition for a New Foreign Policy, a conglomeration of labor groups and religious organizations that lobbied for nuclear disarmament and peace. Rhoads ran the coalition for the next six years, but by 1980, he longed for a change of scenery. Ronald Reagan had just won the presidential election, and Rhoads worried how causes dear to him—environmentalism and antipoverty work and consumer rights—would weather the coming administration. He was aware of how Reagan was openly hostile to those issues and the expenditure of government money on them. So Rhoads regrouped and began to consider new ways to organize for change. Ira Arlook's invitation to work for OPIC in Cincinnati was, to Rhoads, the perfect solution.[16]

Rhoads and his wife moved to Cincinnati in 1980. There, he launched the campaign against construction-work-in-progress. Wanting to grow grassroots support for the issue, he started with a door-to-door canvassing strategy. With anywhere from six to thirty staffers working under him, Rhoads sent foot troops (many of them college students) to the city's different neighborhoods. Knocking on doors, they provided pamphlets to anyone who was interested, asked for donations or membership pledges to OPIC, and helped people call and write their state representatives on the matter. In short, they were trying to make activism easy for people.[17]

To help with the campaign, Rhoads searched for a local activist with experience in organizing and received the recommendation of a woman named Roxanne Qualls. When Rhoads met her in the early 1980s, she was running her own painting and home renovation company (a precursor to her later realtor business). Like Rhoads and Arlook, she was a tenacious activist. Colleagues knew her to be well spoken, even-tempered, and reasonable in judgment. Born in Washington state, she spent her childhood in different locations since her father was in the military. They had a considerable stint in northern Kentucky, though, and from there she attended the University of Cincinnati's College of Design, Architecture, Art, and Planning. Much like Rhoads and Arlook, Qualls's young adult years aligned with the height of the Vietnam War, and, like them, she opposed it. As such, she canvassed for antiwar Democratic candidate George McGovern in 1972 and attended the mass antiwar march on Washington, DC, in 1973 following Nixon's second inauguration. Back in Cincinnati, influenced by the outpouring of women's rights activism across the US, she became the first director of northern Kentucky's Women's Crisis Center in the mid-1970s, when she was only in her early twenties. She also helped direct Women Helping Women, a rape crisis center and shelter for survivors of domestic violence. On top of this work, she was also a staunch environmentalist and consumer rights activist, not to mention a business entrepreneur with her home renovation company. When asked by Rhoads to join the OPIC team, she accepted and began running its donation drives and letter-writing campaign against construction-work-in-progress.[18]

By the end of 1981, when the NRC fined CG&E $200,000 and when the utility filed for another rate increase, OPIC had collected 65,000 local signatures against CG&E's rate requests and its inclusion of Zimmer's construction costs. Then, CG&E had 583,000 electric customers, so over 10 percent of its customer base, at the minimum, stood opposed to its nuclear project for raising their bills. OPIC's first phase against construction-work-in-progress—building local momentum—was clearly working. In the opinion of Rhoads and Qualls, a major reason why they were successful in getting people to care was that OPIC *only* focused on the money. Rhoads, Qualls, and their staffers put aside any personal feelings they may have had about nuclear power (safety concerns, for instance) and instead highlighted to CG&E customers the specific situation in front of them: they were paying for the construction of an overbudget power plant that required expensive construction fixes. Such tact appealed to ratepayers and enabled OPIC to collaborate with local labor unions, including the International Brotherhood of Electric Workers, which had many electricians at Zimmer. Many supported nuclear power for its job creation but could be persuaded to oppose how a particular power plant was financially hurting a community. Years after Zimmer, Qualls said as much: "OPIC's financial niche helped to mobilize more people to reach a critical mass."[19]

OPIC, like Bill Spratley's Office of Consumers' Counsel, and like LaVerne Willsey's Citywide Coalition for Utility Reform, came about in the 1970s and early 1980s, when *consumer rights* was a recognizable phrase and idea to many Americans. This started during the post–World War II economic boom when Americans' spending power dramatically expanded. In the 1950s and 1960s, during a heady time of middle-class affluence, the ability to spend became a key part of being an American, to the point that a major goal of the civil rights movement in these years was better and equal access to places of commerce like department stores, restaurants, and amusement parks. Then, as the environmental movement challenged Americans to rethink their quality of life in the 1970s, public interest advocates like Ralph Nader fought to ensure safer, healthier outcomes for Americans when they bought, consumed, or otherwise used something. Nader and his allies fought for a variety of consumer safety protections, from mandated car seatbelts to improved meat inspection. Holding corporations and government agencies to higher standards, they used nonpartisan public interest organizations and law firms to influence the government from outside of it. Nader alone founded more than two dozen of these organizations. So in their public interest and consumer rights campaigns and in their use of the law and lobbying, Bill Spratley, Ira Arlook, Brewster Rhoads, and Roxanne Qualls were all Nader's counterparts.[20]

In response to the outpouring of interest in consumer safety, presidential administrations from Kennedy to Carter passed dozens of federal laws and regulations to protect people from certain foods, unsafe manufactured goods, dangerous pharmaceuticals, cigarettes, automobiles, toxic chemicals, misleading packaging and advertising, and discriminatory banks and credit agencies. Workers' safety was a piece of this, underscored by the passage of the Federal Coal Mine Health and Safety Act in 1969 and the Occupational Safety and Health Act the following year. And while Reagan's administration, in its attempt to make government small and efficient, attacked these regulatory programs—ordering hundreds of regulations to be postponed, rescinded, or weakened—many Americans by the 1980s had come to expect consumer and workplace protections, guaranteed by the government. And following two energy crises and a recessed economy, fair utility practices were a part of consumer rights to ratepaying households.[21]

In the 1980s, many utility customers used the term *rate shock* to describe severe price hikes that occurred as soon as the costs of newly completed power plants entered the rate base. Across the US, regulators in 1981 approved rate increases that added $8 billion to customers' bills, four times the amount in 1977, 1978, and 1979. As such, local protest emerged against power providers. And as for the companies building nuclear power plants, they in particular encountered ratepaying activists who insisted customers shouldn't pay for construction, given how costly and behind schedule so many nuclear power stations were. Indeed, organizations like the Citywide Coalition and OPIC were far from unique. When, for instance, the Public Service Company of New Hampshire tried to include construction costs from its massively overbudget Seabrook nuclear power plant in the rate base (one of its reactors was set to cost $5.8 billion in 1983), the utility faced such organized vitriol that the state of New Hampshire disallowed the construction-work-in-progress provision in 1979—an outcome that undoubtedly buoyed the hopes of Brewster Rhoads and others at OPIC.[22]

Similarly, state legislatures across America realized that ratepayers had difficulty contesting claims made by well-financed power companies, so—like in Ohio—they, too, created new offices of consumers' counsel or lodged consumer advocates within the offices of attorney generals. There were enough of them that in 1979, Bill Spratley founded the National Association of State Utility Consumer Advocates, the professional organization for state consumer counsels. Indiana's first Utility Consumer Counselor, L. Parvin Price, was very much Spratley's equivalent: he went to battle against the Indiana Public Service Commission, arguing that the utility Public Service Indiana did not deserve rate increases for its expensive and troubled Marble Hill nuclear power plant.[23]

At the same time, congressional representatives from states with nuclear power plants tried to adjust the Price-Anderson Act, which limited utilities' liability in the event of a nuclear power plant accident. Dennis Eckart, a Democrat from Ohio, introduced a state bill requiring Ohio utilities to carry enough insurance to be able to pay for any accident. As a recap, the federal act, set up in 1957, required utilities to maintain a certain amount of private insurance to cover an accident. If damages exceeded that coverage, the federal government's Price-Anderson fund—which was maintained by set contributions from utilities—would cover the rest. But after Three Mile Island, when it became clearer how much a nuclear accident could cost to clean up, people were concerned that private insurance and reserves from the Price-Anderson fund would not adequately cover the entirety of damage. (Three Mile Island's cleanup, after all, was adding up to $1 billion.) Eckart and other utility reform advocates worried that taxpayers might have to bear the rest of the burden, essentially bailing out a utility. Facing the prospect of insurance reform, power company executives balked at the idea, arguing that they could not afford such massive premiums. And while the Price-Anderson Act remained largely unchanged, it was increasingly clear that the utility industry no longer held the power, the unquestioned monopoly, it once had.[24]

While LaVerne Willsey assembled people into city hall chambers in the early 1980s, and as Bill Spratley stood in front of the Public Utilities commissioners in Columbus, and as Brewster Rhoads and Roxanne Qualls walked the streets of Cincinnati collecting signatures, no licensing hearings for Zimmer were occurring—that is, until Wednesday, March 4, 1981, at Cincinnati's downtown federal courthouse. It was the first licensing hearing for the plant since 1979. And as was the case with Willsey's church meetings and the PUCO sessions and OPIC's canvassing, CG&E's money was the topic at hand. That Wednesday, a junior NRC staff member, Michael Karowitz, stood before an Atomic Safety and Licensing Board, detailing his agency's financial projections to finish, operate, and dismantle Zimmer. CG&E executives and their lawyer, Washington, DC–based Troy Connor, along with directors of Columbus and Southern Ohio Electric and Dayton Power and Light, sat nearby, interjecting as needed to back up Karowitz.[25]

The NRC staffer said that it would cost Cincinnati Gas and Electric and its partner utilities another $30 million to finish Zimmer, namely for new safety-related equipment and technical support required after the Three Mile Island accident. He specified that in 1983, when CG&E expected Zimmer to start operating, the plant would cost around $250 million to run and that those costs would climb to $260 million by 1987. Karowitz further elaborated that the three owners of Zimmer planned to "recover all costs of operation through revenues derived from their customers in their system-wide sales of electricity." Additionally, Karowitz explained, the NRC and CG&E expected the Public Utilities Commission of Ohio to continue granting timely rate increases. And why not? CG&E had been successful in procuring rate hikes to date. The expectation that the future would bring more electric and gas sales and adequate rate base amounts was, in Karowitz's words, "a reasonable financial plan" for Zimmer's operation. He concluded by pointing out that CG&E had performed well on its coal plants and that each of the three utilities involved in Zimmer maintained good stock and bond ratings. All of that, he said, should be taken as a sign of future success for Zimmer.[26]

These were the old ingredients of the utility industry: the assumption that state regulators would let a power company include the majority of incurred costs in its rate base so as to enjoy a reasonable rate of return and the assumption that, while the company continued to expend more costs (from money raised through the selling of stocks and bonds), demand would increase, making additional and updated power plants necessary.

At the hearing, intervenors attacked these ideas, arguing that the owners of Zimmer and officials from the NRC should be considering how unexpected financial situations and the current economic climate would impact the nuclear power plant. (By this time, intervenors included the Zimmer Area Citizens groups that had received legal standing after Three Mile Island, their "specific interest" being their residential proximity to Zimmer.)

James Feldman, the twenty-nine-year-old attorney for the Miami Valley Power Project, asked Karowitz, "You have not looked into the possibility such as the Board [the Atomic Safety and Licensing Board] has asked you to look into of there being a Three Mile Island type accident at this facility and the financial consequences thereof? Is that true?" Karowitz responded in the affirmative: "Because such an event is entirely speculative, yes." Feldman continued to press the point. He wanted to know how CG&E would obtain funds for a remediation at Zimmer if insurance only covered a portion of the costs. Earl A. Borgmann, senior vice president for engineering at CG&E, told Feldman that his company and its partners could absorb the costs of an accident at Zimmer "without insolvency." Karowitz added that the NRC considered long-term shutdown costs (after an accident) to be "costs of doing business" and were usually recoverable through a power company's rate base. He further clarified that his commission didn't require license applicants like CG&E to submit financial information regarding their ability to withstand costs of accidents.[27]

Andrew Dennison, the attorney for Zimmer Area Citizens, tried a different line of questioning. He asked Karowitz about rate hikes: What if the PUCO didn't continue to grant them to CG&E as needed? He noted that CG&E and the other two power companies maintained small margins between payments to investors and their net incomes, which suggested that they would struggle to finish and operate an expensive plant like Zimmer without continual rate increases. Dennison pointed out how the PUCO had recently given both Dayton Power and Light and Columbus and Southern Ohio Electric emergency rate increases for fear that the companies would fail without them. To all of this, Karowitz did concede that it would be "extraordinary" for CG&E to survive without timely rate hikes.[28]

Intervenors also asked about the costs and logistics of Zimmer's spent fuel rods. CG&E's original plan was to send them to a low-level disposal facility at either West Valley, New York, or Barnwell, South Carolina. But West Valley closed in 1976 after it leaked high quantities of radioactive gaseous and liquid waste, and the Barnwell site (which had the highest disposal fees in the country) reduced the amount of material it was accepting in the late 1970s, further complicating matters for CG&E. Then, in 1980, Congress passed the Low-Level Radioactive Waste Policy Act, which placed the responsibility of commercial low-level nuclear waste with states. States had to create compacts whereby members collectively decided where, within their member states, their nuclear waste would go. Only in 1985 was the Midwest compact (composed of Indiana, Iowa, Minnesota, Missouri, Ohio, and Wisconsin) finalized. Prior to that, when Zimmer was being built, CG&E had few options for where to send its waste. And because of that—intervenors said—it was difficult to precisely estimate how much waste disposal would cost.[29]

Intervenors also spotlighted the financial uncertainty around Zimmer's decommissioning, set to occur in 2016. The utilities planned to pay for the plant's retirement in the same way they would pay for its operation: relying on recurring rate increases, profits from electricity customers, and continuing investment from bond- and stockholders. And while Cincinnati Gas and Electric estimated the dismantling to cost around $36 million in 1983 dollars, that number came from mid-1970s modeling. Even CG&E leaders conceded that, accounting for inflation, the costs to dismantle might be closer to $41–$43 million in 1983 dollars. The NRC further added that costs could climb to $65 million (in 1983 dollars). But CG&E officials said not to worry. Their company's investment returns would rise "in harmony" with inflation, as William H. Zimmer Jr., senior vice president for finance, told the *Enquirer*. CG&E, in short, was banking on a healthy investment appetite from bondholders to fund Zimmer's decommissioning.[30]

Intervenors picked this apart. It was the case that for new bondholders, CG&E would have to pay them higher interest payments on newly issued bonds to match inflation. But as the hearing made clear, CG&E officials assumed that since these investors would be compensated for inflation with higher payments, they would still invest. So, even with inflation—CG&E's logic went—its fund would be able to cover the decommissioning and avoid a disastrous debt. Company leaders also said they would "fine tune" their fund over the years to make sure it covered costs fully (it was unclear what that meant exactly). Intervenors remarked that higher interest payments to new bondholders were nothing to write off and that it was faulty logic to assume investment interest would keep pace during a recession.[31]

David Fankhauser raised the biggest question of all: Was there enough demand to pay for and justify Zimmer anymore? Since CG&E planned to use revenue from electricity customers to pay for the plant's operation and dismantlement, Fankhauser asked: What if demand and sales were lower than CG&E anticipated? The company's lawyer quickly retorted, "The present demand for Zimmer has already been established at the construction permit stage." But Fankhauser questioned this logic. The Atomic Energy Commission issued CG&E's construction permit in 1972—almost ten years ago at that point—and Fankhauser maintained that electric demand had undoubtedly dropped since then due to the energy crises and poor economy of the 1970s. By 1980, inflation was close to 15 percent. Electricity customers thought twice before flipping on the light switch.[32]

Science writer Eliot Marshall captured this sentiment in 1981 when he announced that "America's electric utilities are in trouble." Marshall, a health, medicine, and science reporter, described a nationwide problem with the leaders of the electric industry, arguing they had misplaced ideas of demand: "They are losing income because the demand for electricity has fallen, construction costs are rising rapidly, and the public is refusing to pay higher electric bills." The "high-growth assumptions that have guided the industry for the last 20 years," Marshall said, were becoming obsolete. (Fankhauser brought a copy of this article with him to the hearing.)[33]

In the same way that the 1973–1974 recession lowered customer demand and cast doubt on the financial sense of commercial nuclear power, the recession of the early 1980s did as well—even more so, in fact, given how much utilities like CG&E had sunk into nuclear power plants by then and how much the economy had worsened over the 1970s. Crippling inflation continued to make labor and materials, like Zimmer's parts and fuel, cost-prohibitive. To this point, in 1971, the Atomic Energy Commission had said that capital costs to build a nuclear station should not exceed $150/kilowatt, but by 1976, the Nuclear Regulatory Commission had already increased that estimate to $1,200/kilowatt. Compounding utilities' troubles, bond rating companies, including Standard and Poor's, downgraded power companies in the late 1970s and early 1980s to encourage them to cut back on expensive construction projects. Downgraded ratings disincentivized people from purchasing bonds from utilities since investing appeared riskier. Ongoing issues with coal and gas also added to utilities' financial troubles: anti–air pollution upgrades were still proving very costly; meanwhile, inflation and labor agitation drove up coal prices in the late 1970s. All of these financial issues—combined with other chronic ones, like long regulatory delays and construction issues that required expensive fixes—made nuclear power an increasingly unlikely prospect in America.[34]

This was manifest in the high price tags attached to nuclear power plants by the early 1980s. The Shoreham plant in New Hampshire, which planners had expected to build for $75 million when construction started in 1967, cost $2 billion by 1982 and was, like Zimmer, not yet operable. The utility in charge was then paying $1.5 million *a day* in holding costs alone. On the opposite side of the country, the owners of the Diablo Canyon nuclear power station had invested $2.2 billion into their plant by the end of 1981, about 18 percent of the company's assets. This plant was also inoperable. North of there, the Washington Public Power Supply System's second nuclear power plant in Hanford, Washington, was supposed to be finished by 1977 at a cost of $352 million. In 1980, the cost had risen to $1.82 billion, which included $6 million per month in holding costs. At the end of the year, only one out of the owner's six planned nuclear power stations was in operation, and the five others in progress had seen construction costs go from $4 billion to more than $17 billion. These dire financial situations existed across the nation, and they took the allure out of nuclear power for a growing number of utilities. By March 1979, just *before* the Three Mile Island accident, power companies had canceled fifty-two ordered-but-not-yet-built nuclear power plants. One year before that, CG&E canceled the second reactor unit planned for Zimmer, telling shareholders, "Nuclear generation still is economically attractive, but less than the 1969 studies indicated."[35]

So when Fankhauser asked if there was enough demand to pay for and justify Zimmer anymore, he was acknowledging all these economic and industry developments. CG&E's attorney objected to the line of questioning, claiming that national trends concerning the energy industry were irrelevant to Zimmer. But this time, the Atomic Safety and Licensing Board agreed with Fankhauser. Its members insisted that widespread problems for the utility industry *did* have a bearing on Zimmer. (In this way, the board was operating as it was supposed to, providing an independent check to the NRC and to the licensing process.)[36]

But CG&E executives and lawyers did not answer Fankhauser's question in any meaningful way. In fact, not once, in any press interview, in any licensing hearing, or in any stockholder report, did they concede that they had a hand in creating any of the issues confronting them. Instead, company leaders maintained that the Cincinnati area still needed Zimmer and that the plant's high costs were due to unnecessary regulation, bureaucratic delays, inflation, and labor disputes—all of which was, in their opinion, out of their control. "The economic and environmental benefits of nuclear power are well established," they told shareholders in 1980. "The point seems clear to leaders of our scientific and technical communities, but has not been grasped by the general public and government policy makers. The unbelievable complexities of governmental regulations, many of them contradictory, have ballooned the costs and construction times of nuclear power plants."[37]

While it was true that some issues, like inflation and the length of time it took federal and state agencies to analyze a nuclear power plant permit, were out of utilities' control, many factors were not. Inexperienced and negligent management (as whistleblowers exposed) and a certain arrogance among utility management were major causes behind the extensive regulation, bureaucratic delays, labor disputes, and mounting costs with nuclear power. Furthermore, while CG&E and its partner utilities never stated so publicly, they had a strong incentive to complete their expensive nuclear power station, even as it cost more and more: they had already invested $1 billion into it, and they could only earn a return on their money if the plant operated.[38]

Contributing to CG&E's intransigence over Zimmer—its unwillingness to admit any wrong or consider canceling the project—was that its executives, like those in charge of other power companies, were enduring unsettling competition in their industry following the energy crises. In 1978, the Public Utility Regulatory Policies Act was an unprecedented promotion of energy conservation and the nontraditional energy market of renewable energy and cogeneration (cogeneration is the procurement of two forms of energy from one fuel source, like electrical and thermal energy being generated from a single natural gas turbine). The act permitted alternative energy providers (i.e., not traditional power companies but rather those in renewables and cogeneration) to operate unregulated and to fetch fair prices for the energy they produced (which utilities then bought). Because of these factors, and because nontraditional energy producers proved adept at technological innovation, they began to offer comparable prices to traditional utilities. This, in turn, began to disrupt the monopoly power companies had held since the 1800s. (It started the movement toward energy deregulation and market competition that would come to many states in the 1990s and 2000s.)[39]

These changes profoundly disquieted the men in charge of utility companies. So did the rising ethos of energy efficiency and conservation among government officials, academics, and some sectors of the public by 1980, and not just because it was "the environmental thing" to do. Policymakers were encouraging utility companies to focus on the demand side of their business—actual demand, not invented—instead of the supply side as they had traditionally done. And as a part of that, government officials suggested that smaller-scale generation and a lower growth rate of electricity consumption would be more efficient, would more accurately meet demand, and—for utilities and customers—would be less costly than building a bunch of large power plants (like nuclear ones). These analysts also questioned the utility industry's long-standing claim that energy development stimulated the economy.[40]

To these radical ideas and changes, many utility managers dug in their heels. They argued that conservation was a short-term fix during a period of high energy prices—and that as soon as the oil crisis ended, things would return to "normal," which for them meant building traditional power plants without interference. Furthermore, utility executives did not want to pursue conservation for the simple reality that reduced sales meant lost revenues for them. Summing all this up, *Fortune Magazine* described the leaders of power companies in 1980 as "generally unimaginative men, grown complacent on private monopoly and regulated profits."[41]

Back at the 1981 hearing, as intervenors and CG&E counsel dueled over Zimmer's financial future and the dynamic state of US energy, the NRC representative Karowitz was evasive in giving the Atomic Safety and Licensing Board his expert opinion on CG&E's financial readiness. "It's very difficult," he told the board, "to quantify with any degree of precision over a long period of time. The [NRC] review does not contemplate an extremely refined analysis to take place, but a general assessment is made of the relative capabilities of the Applicant." When one of the intervenors asked Karowitz if the NRC had taken financial data and assessments directly from CG&E, without additional NRC review, Karowitz replied, "That is correct." At this point, the chairman of the Atomic Safety and Licensing Board stopped the hearing and asked Karowitz what was really being accomplished then. "Well, what we are really missing," the chairman said to the NRC official, "is any trace of your analysis at all of this data. There is nothing for us to see." He asked Karowitz what his qualifications were. Karowitz replied that he was a graduate student. The chairman threw up his hands, exasperated that the NRC hadn't sent (in his mind) a qualified representative.[42]

While he was a junior employee, Karowitz wasn't wrong in how he represented the NRC. In the late 1970s, the commission relaxed its requirements for proof of financial preparedness for utilities seeking to operate nuclear power plants. According to the updated rules, CG&E needed to prove only a "reasonable assurance of obtaining the necessary funds" to operate Zimmer. The NRC also specified that "a utility cannot provide more than a reasonable assurance that funds will be available through the course of a multiyear construction project." The commission recognized that numerous factors—like stock and bond markets, changing regulations, and cost of fuel—made it hard for a utility to put one budget in its license application and stick with original projections. On the one hand, such factors *did* make future projections difficult, yet intervenors rightfully called out how CG&E had not put significant work into at least estimating different financial situations (i.e., investment returns not rising with inflation, or the PUCO not granting timely rate hikes). Instead, intervening parties felt yet again the nagging sense that the NRC was not doing its job well. Regurgitating CG&E financial assessments and asking for only a "reasonable assurance" of financial preparedness made the commission seem an apologist for utilities, not a regulator seeking to capture the public's confidence.[43]

The hearing ended inconclusively, but one year later, in March 1982, the NRC eliminated case-by-case financial review for applicants seeking a nuclear power plant operating license. The NRC argued that this review, especially during the construction stage, prompted many utilities—which were dealing with the same economic and financial troubles as CG&E—to abandon their nuclear power plant projects. Furthermore, commission officials stated that they found no association between financial preparedness and safety outcomes for utilities and their nuclear power stations.[44]

The decision not only rendered the Zimmer hearing in March 1981—all of the time and energy that intervenors had put into preparing for a technical, financial discussion—meaningless, but it also reinforced to Fankhauser and his allies that the NRC, just like the Atomic Energy Commission, was more interested in promotional work than protecting utility customers and public safety. The reversal in policy was yet another strike against the NRC.

One day in the fall of 1981, Brewster Rhoads from the Ohio Public Interest Campaign got a phone call from a former worker at Zimmer. The worker did not feel comfortable identifying himself, but he had read about Rhoads in the local newspapers, about his work against the construction-work-in-progress provision, and asked to meet with him over some concerns he had about the power plant. Rhoads met him in the parking lot of a Frisch's Big Boy in Fairfax, a tree-lined working-class suburb to the east of downtown Cincinnati. After much prodding, Rhoads convinced him to come inside, where they found a discreet table in the corner. Sitting there, the man—shaking and constantly looking over his shoulder—finally shared his name. He was Ron Yates, a Cincinnati resident, and he had been a journeyman plumber at Zimmer from late 1975 to May 1981. Warming up to Rhoads, he disclosed that he had witnessed welders performing faulty work throughout the plant and that inspectors weren't noticing it in their assessments. Fearful that the welds would not hold, Yates shared that he didn't know what else to do other than blow the whistle. With their first meeting, Rhoads earned Yates's trust, and the two started to regularly meet. Zimmer had produced yet another whistleblower.[45]

Then, from Wednesday, December 9, 1981, to that Friday—a few weeks after the NRC issued its $200,000 fine—there were public PUCO hearings in Columbus, Ohio. The topic at hand was CG&E's recent request for a $134 million electric rate increase. Rhoads thought this would be a perfect opportunity for Yates to come forward with his information, knowing it would complement what Tom Devine had already discovered about CG&E and NRC mismanagement. Rhoads contacted Bill Spratley and his Office of Consumers' Counsel, who agreed to provide legal counsel to Yates at the hearings. There, the plumber went public, sharing multiple instances of poor and wasteful work during his time at Zimmer. Yates explained that once he spent thirty days reinstalling pipes that had not been installed correctly in the first place. And when workers couldn't locate materials they needed—materials, Yates explained, that were mandatory parts for safety-related construction—they simply carried on without them. The whistleblower also described how workers stole tools from the site, and because things were so disorganized, men stood around for up to an hour, waiting for the tool they needed to become available. Similarly, he recalled being hired for stretches of time and then not being given any work: "I spent many a day standing around wondering why they hired me."[46]

The Office of Consumers' Counsel then took the floor, telling the Public Utilities Commission, yet again, that ratepayers shouldn't be paying for Zimmer's seemingly bungled construction. "It's time to stop bailing the utilities out," Bill Spratley said. Noting that Zimmer's costs had by then escalated to $1.3 billion, Spratley then called for an independent audit of Zimmer by the PUCO. Customers should not pay for costs "which are a direct result of management negligence or imprudence," he told the commissioners. Spratley also asked the PUCO to deny CG&E the rate request. Company officials, unsurprisingly, started at both demands. One CG&E representative said an audit would be "wheel-spinning at taxpayer expense."[47]

A little over a month later, on Wednesday, January 27, 1982, everyone received the verdict: the PUCO was granting CG&E an annual electric revenue increase of $85.4 million and an annual gas revenue increase of $19.4 million. It was less than CG&E asked for, but the PUCO did not seem to be pursuing any kind of audit, and moreover it was permitting 50 percent of the utility's investment in Zimmer as of March 31, 1981, to factor into the rate base.[48]

By the following year, CG&E ratepayers had paid $70 million for Zimmer's construction through their rate base and Columbus and Southern and Dayton Power customers another $115 million. Because of this—and because of new and major safety issues—1982 would prove momentous for the fight against Zimmer. A major coalition against the plant emerged then, and its focus was on safe *and* affordable energy. It was the moment when all the earlier strands of protest came together in a giant crescendo.

7

"It Was Like a Dam Breaking"

A Coalition against Zimmer

On Wednesday, February 24, 1982, CG&E officials paid their $200,000 fine to the NRC. Around the same time, licensing hearings convened to discuss CG&E's preparedness for an accident at Zimmer—the very last topic that required licensing board approval before the NRC could grant CG&E an operational permit. Alongside the hearings, CG&E staged mock accidents at the plant, required by federal officials, to show how the utility and local, county, and state agencies would respond in the event of a situation there.[1]

None of it went well for CG&E. The Federal Emergency Management Administration (FEMA) failed the utility on emergency responsiveness, and on Thursday, June 21, 1982, an Atomic Safety and Licensing Board agreed with FEMA. Before the utility could get its final license for Zimmer, it would have to redo its emergency plans and stage new accidents, which would have to have more successful outcomes. Groups and people organizing against Zimmer, shocked at the inadequacy of accident planning, cheered at CG&E's failing marks. Indeed, even groups like Zimmer Area Citizens, which were previously invested in making Zimmer safe for the communities around it, had been radicalized by the plant's worsening state: they, too, wanted the project canceled.

From CG&E's perspective, the rest of 1982 proceeded similarly, and Zimmer's prospects—despite the plant being 95 percent complete—continued to dim. As mandated by the NRC as part of the $200,000 penalty, the utility launched a massive reinspection of Zimmer, called a *quality confirmation program*. But, instead of resolving issues for CG&E, the program revealed more problems, in large part thanks to the relentless work of Government Accountability Project attorney Tom Devine. Concerned that the quality confirmation program would not fix Zimmer, since the NRC had placed CG&E and its general contractor Kaiser in charge of the program, and concerned that the utility and Kaiser would sacrifice safety for a speedy audit (considering the plant already cost $1.25 billion), Devine learned from whistleblowers that this was, in fact, the case: the program, which lasted from November 1981 to November 1982, was frighteningly cursory. Not only did upper management harass inspectors, but Devine further found that both CG&E and sections of the NRC had been criminally negligent in quality assurance work in the past. These revelations coerced the NRC into unprecedented corrective action, including a safety-related construction shutdown at Zimmer.

In the midst of this, Devine and Tom Carpenter agreed it was time to unite all the groups and people opposed to—or even somewhat critical of—Zimmer's costs *and* construction issues. In the summer of 1982, the Coalition for Affordable and Safe Energy formed as a broad-based effort to coordinate anti-Zimmer opposition. In its sole focus to end Zimmer as an unsafe and costly power plant (note: power plant, not a *nuclear* power plant), it represented a fresh take on the Zimmer debacle: a pragmatism that aimed to secure a realistic outcome and unite a lot of people while doing so. And yet in its values around functioning democracy and accountable leadership, the coalition was a continuation of all that came before it. Under its platform, leaders gathered fifty organizations and an impressive twenty thousand people. In what otherwise looked like a group of odd bedfellows, CASE—as the coalition was called—ingeniously included anti-Zimmer groups, intervenors, CG&E customers, city of Cincinnati representatives, local construction workers, Zimmer whistleblowers, coal and railroad miners, environmental and peace activists, consumer rights groups, and, with the involvement of local civil rights leader Marian Spencer, Black Americans from around Cincinnati. Individuals and groups as distant as eastern Kentucky, West Virginia, and central Ohio also joined. CASE's leaders and rank-and-file organized alongside Devine, providing widespread grassroots support to his legal action. As one coalition leader later recalled, it felt like a crescendo of energy against Zimmer—or, in his words, "it was like a dam breaking."

After Three Mile Island, the NRC instituted a new framework for accident consideration, requiring that for a license, a utility had to provide "reasonable assurance that adequate protective measures can and will be taken in the event of a radiological emergency." While vague in language, the provision forced power companies, with the assistance of state, county, and local authorities, to plan for residents' safety in the ten-mile radius around a nuclear plant. For CG&E, that meant accommodating twenty-five thousand residents. The utility had to get its plans approved by an Atomic Safety and Licensing Board and the new Federal Emergency Management Agency, created by an executive order of Carter's in 1979. FEMA, in particular, had to approve or deny CG&E's performance during an accident simulation.[2]

For the drill—like for an actual accident—CG&E had to work with a plethora of entities at all levels of government: FEMA; state organizations (from the governor's office to Ohio and Kentucky National Guards); county organizations (like disaster relief agencies from the counties near Zimmer); local organizations from Ohio and Kentucky (from local police, fire departments, and emergency medical services teams to city, township, town, and village officials); and volunteers from Red

Cross and other such groups. Emergency preparedness entailed massive coordination, and even though all parties involved knew about and planned for the simulation ahead of time, it was soon clear that CG&E and its partners in disaster relief were not prepared for a real accident.

The first drill began at 7:27 a.m. on November 18, 1981, a sunny and cold Wednesday morning. Stone and Webster, a New York–based engineering firm hired by CG&E (a firm with a long history of working in both nuclear weapons and commercial nuclear energy), informed operators at Zimmer that reactor core coolant had leaked from the primary containment structure. By 7:44 a.m., the reactor was shut down. Throughout the morning, CG&E staff at Zimmer notified nearby police forces and evacuated personnel from the reactor building. After a "significant" amount of radiation had swept over sections of southwest Ohio and northern Kentucky, Ohio governor James Rhodes declared a state and then a general emergency. Local police and fire departments, EMS teams, Red Cross volunteers, and the Ohio National Guard mobilized as if radiation had contaminated the area surrounding Zimmer. Some of these responders were dispatched to Batavia High School, about twenty miles north of Zimmer, to set up a decontamination center, where high school students pretended to have been irradiated. During the mock accident, a small group of protesters spent the day at Point Pleasant Park near Zimmer, making sure CG&E knew they were not reassured by the drill. David Fankhauser and his young daughter held signs that read "Health before Profits."[3]

By the end of the day, it was apparent the simulation hadn't gone well. CG&E executives complained that the drill was more stringent than they had expected. One of them, Earl Borgmann, told the press, "We had a very tough day." As a part of the test, eight major systems at Zimmer had failed, including the reactor cooling system, which CG&E officials had not anticipated being a part of the trial. Nonetheless, they remained undiscouraged. The company's new vice president for nuclear operations,

B. Ralph Sylvia, commented at the time that such a scenario was very unlikely to happen. A more critical response came from the police chief of New Richmond, close to Zimmer. He told local reporters, "Thank god this was only a drill." Among other problems, he noted that when evacuation procedures were being tested, a school bus full of children had driven *straight into* the mock radiation zone, indicative that departure routes and coordination were unclear. The police chief stated afterward, "We definitely had problems. Others might not say anything, but, doggone it, I've got to live in this area and I was not convinced we would have been able to handle a real problem."[4]

In June 1982, another trial occurred. It had issues, too, communication and public relations being two of them. In an era before mobile phones, responders relied on landline phones and radio to communicate to each other, but the staged accident showed that Clermont County, where Zimmer sat, had major radio dead spots that would not receive a warning. Phones lines also became overwhelmed. Officials from Clermont County Disaster Services Agency had tried to phone the superintendent of a local school district, as the drill required, to let them know of the accident. But CG&E management had failed to notify the local community of what was going on, so when sirens sounded in real life, unknowing parents were alarmed and called the schools in mass, worried about their children. The number of phone calls overloaded the system and disabled the lines. The exercise additionally revealed that evacuation was still a problem, particularly for children, the elderly, and those physically disabled. Both staged accidents made clear that there were far too few school buses for a swift evacuation.[5]

Shouldering much of the accident response, local officials from the towns near Zimmer expressed concern that most emergency services were run by volunteers. This made them wary of the outcome of an actual accident at Zimmer. The mayor of Mentor, Kentucky, told the press his concerns: How reliable and available would *volunteers* be? When would they be trained? How would their own personal welfare or that of their family conflict with their responsibilities? The mayor also noted that Mentor, home to about two hundred people, had a budget of only one hundred dollars for its emergency planning.[6]

The issues that plagued the accident trials—poor communication, an unclear chain of command, a reliance on volunteers, and not enough funding—reflected the national state of incidence management. But it was also the case that CG&E officials had not tried to address these gaps in emergency management. That was because CG&E officials did not believe an accident of significance was possible. This was made obvious at Atomic Safety and Licensing Board hearings, convened prior to CG&E's second simulated accident.

There, the topic was CG&E's emergency preparedness, and intervenors came prepared, noting issues from the first staged accident and anticipating many of the problems that would occur at the second. Members from the Zimmer Area Citizens groups spearheaded this effort. They came with a sixty-three-page itemized list, detailing errors and blind spots in CG&E's emergency planning. In this role, they made it known that they had changed their minds on Zimmer. Unlike in 1979, when they suggested better emergency planning for Zimmer, in 1982, they were convinced that the issues at Zimmer were too numerous and CG&E's management too poor: the plant was unsavable.[7]

Members of Zimmer Area Citizens argued that the overarching problem was that no one—not CG&E, not state, county, or local officials—had a clear and coordinated response plan, and CG&E was not taking the initiative to address the shortcomings. For instance, CG&E planned to alert Ohio and Kentucky county authorities first, as soon as an emergency happened, and then those officials would contact more local authorities to initiate a response. It sounded simple enough—too simple, in fact. The attorney for Zimmer Area Citizens, Andrew Dennison, said that the cursory plan would result in muddled lines of authority and communication. He also pointed out how evacuation routes ended at Clermont County lines. Technically, CG&E was following rules: nearby Brown County, for instance, was outside the ten-mile radius. But Dennison noted how it was only fifteen miles east of Zimmer and had much wider roads than Clermont County, making any plan that ignored it seem unwise. Timeframes between different response teams were off as well, with everyone disagreeing over what constituted a "reasonable time" for evacuation. Clermont County Disaster Services estimated that it would take forty-nine hours to evacuate all twenty-five thousand residents within ten miles of Zimmer. CG&E argued it would take a little over four hours. Zimmer Area Citizens had approximated around seventy-seven hours.[8]

Another issue was that counties were supposed to select relocation shelters for people in the ten-mile zone around Zimmer, and CG&E was supposed to identify routes to get people there, but Dennison noted that the utility had used old, inaccurate maps to create evacuation pathways. Furthermore, the utility did not choose particularly wide streets to accommodate increased traffic. These decisions meant people would not get to county relocation centers easily (if at all). Dennison also told the Atomic Safety and Licensing Board that evacuation schemes were too simplistic: they had failed to consider human error and emotion. Some people would be flustered. Others would, on learning of an accident, return home to retrieve children or elderly parents, possibly moving closer to the plant to do so. One-direction movement among twenty-five thousand strangers could not be counted on, Dennison said.[9]

The families who lived around Zimmer understood that the area's rural nature and its narrow, meandering roads made evacuation a major challenge, but they were wholly unsympathetic to CG&E's solutions—inarguably inane ideas that were almost certain to fail. To help spread the message that residents needed to leave, CG&E officials came up with the idea that residents should hang a green cloth on their doorknob to tell neighbors and authorities that they needed assistance in being evacuated. Dennison explained how this plan necessitated people—who were supposed to be evacuating—instead driving down long, private driveways to see if their neighbor needed assistance. Similarly, CG&E's plan called for residents to signal that they had evacuated by placing a towel or sign on their mailbox, letting authorities know they had gotten out of harm's way. CG&E also had not considered how severe weather events like snowstorms and flooding might affect evacuation on rural roads. To this, Zimmer Area Citizens leader Alice Gerdeman—the nun and elementary school principal near Zimmer—told the Atomic Safety and Licensing Board, "You can't tell a nuclear accident it can only happen on a sunny day."[10]

She also pointed the board to the *Circle of Safety* book that CG&E disseminated to Zimmer-area families. It was supposed to be a guide for local families in the event of an accident, but as Gerdeman noted, it failed to give readers a clear picture of what to do. Basic information, like which relocation center to go to for which service, was not included. Dennison called it "unreadable." FEMA had also provided a 269-page book, *Radiological Monitoring, Four Self-Study Units*, given to volunteers training to aid in a nuclear power station crisis. Most of it was dense, unhelpful, and (ironically) focused on how safe nuclear power was. Only the last few pages explained what one should do in a nuclear energy accident. For Zimmer, people were told to go to the nearest military base—Fort Thomas in Kentucky—and the commander there would call the governor for next steps. That was all the information residents were given.[11]

School evacuation was another unresolved issue. Gerdeman explained to the Atomic Safety and Licensing Board that CG&E's plans for schools were both illogical and insensitive. The utility anticipated that principals would receive news of an accident via phone call, yet Gerdeman noted that most principals in budget-strapped schools near Zimmer also taught classes and did not regularly receive phone calls. CG&E then suggested installing additional alarm systems for each school that would sound in the midst of an accident. Gerdeman shook her head at the idea, noting how distressing that would be for schoolchildren to hear.[12]

The issues kept coming. Dennison told the Atomic Safety and Licensing panel that there were not enough school buses to remove all 2,600 students near Zimmer and that not all school buses were fitted with radios. Furthermore, most bus drivers were not full-timers. They were farmers or mothers who returned home after a shift to care for small children. But, Dennison said, even if they were full-time bus drivers, they were not necessarily parked by a school all day, listening to their radios for news from Zimmer. Rather, buses and drivers were at scattered locations during a typical day. He wanted to know how, outside of a radio message or telephone call, bus drivers would learn that an emergency was occurring. Furthermore, what was a bus driver to do if they were transporting students when a decision was made to evacuate? Should that person take students home immediately or assume responsibility of them?[13]

In response to these questions, CG&E's attorney explained that CG&E planned to sound a siren loud enough to attract everyone's attention within the ten-mile radius around Zimmer. He further said that even if bus drivers did not receive word of an accident, they would pick up children as usual in the morning, successfully getting them to the school shelter. At this point, the Atomic Safety and Licensing Board chairman responded incredulously, "Would that seriously be the advice [you give]? That in the middle of a [radioactive] release during the start of the school day, people, instead of taking the protection action of sheltering, go out and catch their bus to school?"[14]

At the hearings, it was patently clear that CG&E officials were dismissing the likelihood and severity of an accident. The utility's lawyer said that on the off chance there was a radiation release and people needed to take shelter for a period of time, one week's supply of rations would suffice. A mother of eleven children who lived near Zimmer stood up and asked him, "So I'll have to have a semi in my yard? Milk for 11 kids? Diapers?" Local officials at the hearings also pointed out that they did not have enough medicine to treat people for thyroid radiation contamination. In response, CG&E's counsel said that local governments should buy more.[15]

The emergency management issues that confronted the communities around Zimmer weren't unique to southwest Ohio. In fact, other places with nuclear power plants were dealing with the very same concerns: too few buses, overreliance on an unsteady supply of volunteer labor, inadequate alternative pathways out of a community, and tight money supplies for emergency resources. And they were dealing with utilities that, like CG&E, tended to treat emergency planning as an exercise in public relations. At licensing hearings, power companies were known to prepare mediocre flow charts, diagrams, and slide shows because they did not believe accidents would happen and because they resented the additional time and money spent on planning for such occasions. On top of that, they often sparred with disaster relief authorities in designing emergency plans because local officials and teams were the ones to find the holes in plans. It was local authorities, after all, who would be at the front line of any accident, so they took the exercise seriously, including admitting that they lacked coordination and resources.[16]

Making the situation more tense, the NRC usually sided with utilities, telling the public that nuclear power plant licensure would not be held up by local authorities unhappy with evacuation routes. Indeed, in Cincinnati, the NRC announced in late April 1982—in the midst of licensing hearings, accident trials, and the NRC's mandated reinspection of Zimmer following its $200,000 fine—that the plant should soon receive its operating license. *Cincinnati Post* writer Ron Liebau captured the stunned local reaction: "The NRC's staff recommendation—which the licensing board may adopt, modify or reject—follows nearly three years of public hearings, hundreds of witnesses and thousands of pages of transcripts. Thirteen days were spent this year on emergency planning."[17]

Around the country, the NRC and the utility industry's dismissive treatment of emergency planning propelled local communities to fight back. At the Shoreham nuclear power plant on

Long Island, residents, along with county and state officials, rallied around the call "You can't evacuate Long Island." Given that the plant was within the metropolitan area of the nation's largest and densest city, local officials and residents were horrified that Long Island Lighting Company had not considered heavy traffic and congested exit routes when it designed egress means for the island. In the same way, botched emergency planning at Zimmer grew the antipathy of Zimmer-area residents and officials against CG&E. Just like how the aftermath of Three Mile Island spurred many of them to question Zimmer's safety, the careless evacuation plans made it painfully clear to those community members—and others—that an accident at Zimmer could be life threatening. City officials from Mentor, Kentucky, became so worried about an accident that they successfully petitioned the NRC to be an official intervenor in Zimmer's remaining licensing hearings.[18]

Fortunately for many communities, approval of emergency planning wasn't just up to the NRC. First, the ball went to FEMA, and the agency stood its ground, withholding approval of emergency planning for more than one nuclear power plant. Cincinnati's was one of these. After the second simulated accident at Zimmer, FEMA officials failed CG&E. As part of the drills, FEMA was supposed to have been notified. That communication, in FEMA's opinion, took far too long to come through. There were other outstanding problems too—all the issues intervenors had raised. FEMA officials informed CG&E management that the utility needed additional and more successful trials before they would give their approval. With FEMA denial, the Atomic Safety and Licensing Board—the second in command to approve or deny emergency plans—followed FEMA's lead. And with that, Zimmer's licensing hearings stopped. CG&E representatives, forced to embark on another design round of emergency planning, announced in great frustration that they would not be ready for another simulated accident until the summer of 1983. NRC staff, ever hopeful, said that licensing hearings might resume by spring of 1984, after which a license would be possible.[19]

Meanwhile, Tom Devine was busy at work. As legal counsel to the Miami Valley Power Project, Tom Carpenter's Cincinnati Alliance for Responsible Energy, and various whistleblowers, Devine closely kept on the trail of CG&E and NRC officials after the $200,000 penalty, when the mandated reinspection of Zimmer—the quality confirmation program—launched late in 1981.

The program started on an interesting (if unsurprising) note, to say the least: CG&E officials disavowed responsibility for most of the violations at Zimmer. Company leaders publicly stated that since their general contractor, Kaiser, had overseen quality assurance work prior to 1980 (after which CG&E took a more direct role), CG&E was ignorant of and blameless for problems that dated to the 1970s. The company should not be the primary one responsible for fixing the mess of errors, they said.[20]

Devine said and showed otherwise. Using the Freedom of Information Act, he obtained NRC records that laid bare how CG&E management had stopped Kaiser's quality control inspections in the 1970s and refused adequate inspection staffing. Devine then reminded the NRC that its own Office of Inspector and Auditor had launched a criminal investigation into CG&E in the summer of 1981 for these actions, trying to determine if the utility had deliberately violated the Atomic Energy Act of 1954. Under the act, it was a federal offense to falsify records at a nuclear facility or harass quality control inspectors. Both had happened at Zimmer, as Devine revealed that summer, but to his frustration, the Office of Inspector and Auditor decided to suspend the investigation.[21]

As he was disproving CG&E's claim about 1970s construction management, Devine noticed something else that was very troubling—this time, about the NRC. When he requested NRC documents through the Freedom of Information Act, certain pages were omitted. When Devine pressed the NRC about the missing documentation, officials from its Office of Inspector and Auditor claimed those records did not exist. But Devine knew they did. He was aware that a whistleblower—an NRC quality assurance inspector—had come forward to other NRC officials with damning evidence against Zimmer's quality assurance program. But the transcript of the inspector's interview was missing from the papers that Devine had obtained. Another whistleblower from the NRC clarified the situation: when they called Devine, they shared that reports from the Office of Inspector and Auditor were being purposefully hidden in the basement of the office's head official.[22]

With that jaw-dropping development, Devine filed suit against the NRC in the US District Court for the District of Columbia, charging that the commission illegally withheld information from the public record. He also submitted his evidence to the FBI and to Congress, which subsequently launched congressional investigations on the NRC and CG&E's management of Zimmer.[23]

One result of this was, on Thursday, June 10, 1982, Devine appeared before a congressional subcommittee headed by Morris Udall, a long-standing Democrat representing Arizona and a legislator with a history of taking antiwar, pro-environmental, and antinuclear positions. Devine understood that it was CG&E, Kaiser, and the NRC's *past* actions—the destruction of the public record, among other things—that had resulted in the congressional inquiry, but he wanted to use his time in front of Udall to instead focus on Zimmer's ongoing quality confirmation program. The past couldn't be changed, after all, and in his professional opinion, the only good he could do for his clients was to convince Udall and other authorities that the NRC's quality confirmation program wouldn't actually fix Zimmer.

Devine explained that CG&E and Kaiser's past actions rendered any hope of conducting a comprehensive review of Zimmer nearly impossible. For instance, the company and its contractor had sourced parts for Zimmer from junkyards, so when it came time to verify the original integrity of those parts, workers could not do so. "Similarly," Devine asked, "how can CG&E determine today that a vendor had a reliable QA [quality assurance] program when a part was purchased 8 years ago?" He noted that this issue—that parts were too old to verify their original quality assurance—affected 42,000 parts inside Zimmer. As such, Devine argued that only a "massive repurchase program"—of parts and equipment—would begin to ensure Zimmer was built well.[24]

But beyond these specific parts issues, the lawyer was convinced the reinspection program wouldn't work because, in his words, it did not "address the causes of the QA breakdown" in the first place: utility mismanagement and inexperience. And because the NRC largely put the quality confirmation program in the hands of CG&E and its general contractor, Devine called the program "basically a paperwork review" and argued it would result in the same disastrous outcome as the original quality assurance work. "In effect," Devine stated, "the NRC told the fox to do a better job of guarding the henhouse. The results have been predictable."[25]

To prove this point, Devine shared stories of the quality confirmation program, as disclosed to him through whistleblowers. For instance, CG&E and Kaiser managers told field inspectors to only recheck items identified by the NRC (not by whistleblowers) and to only report back to CG&E and Kaiser managers, limiting the review process and any outside transparency. Devine—worried that safety issues might go undiscovered, that reports of nonconformance might be destroyed, and that issues brought to light by inspectors might be suppressed by the utility and Kaiser—showed that his concerns weren't just hypothetical. He told Udall's subcommittee, "While CG&E assured the NRC in February [1982] that harassment of Quality Assurance personnel has ended, just two weeks ago three inspectors were doused with 'dirty water.'" In response to the harassment, an NRC inspector commented at the time, "I'll just say it's the same type of thing going on for a while."[26]

More to this point, Devine brought with him two new whistleblowers who shared past and ongoing negligence in quality assurance and continuing harassment of inspectors. One, David Jones, was a quality assurance engineer and manager for Kaiser. Jones divulged that as early as 1969, CG&E and Kaiser management had intentionally avoided as much quality assurance work as possible at Zimmer to build it as quickly and inexpensively as possible. Fast forward to the spring of 1981, when Kaiser assigned Jones to investigate safety-related construction issues at Zimmer raised by the NRC's Office of Inspector and Auditor. When Jones agreed with the office's report (that there were problems at Zimmer), he suddenly found himself removed from the project. On top of that, Devine told the subcommittee, "Mr. Jones suffered the bulk of his harassment after April 1981"—that is, within the last year.[27]

"In our opinion," Devine said to Udall, "the only way to save this nuclear power plant is by eliminating the structural conflict-of-interest. An independent organization—with nothing to lose by finding violations—must replace both the Kaiser and CG&E QA [quality assurance] programs, as well as the QCP [quality confirmation program]." In this way, he maintained his moderate stance toward nuclear power, not condemning it but trying to make it as safe as possible. Devine said that if a third-party audit did not occur, "the cloud of public mistrust in the safety of Zimmer will remain. And the safety defects will remain dormant at the plant. In the long term, those consequences will not benefit anyone."[28]

In the midst of the subcommittee hearings, the US Department of Justice announced a criminal investigation into whether CG&E had deliberately jeopardized Kaiser's quality assurance in the 1970s by refusing the contractor staff and funds for safety inspections. Devine made sure Udall heard the news, telling the subcommittee, "Unless the U.S. Attorney chooses not to pursue those leads, we are now faced with a curious scenario: The same organization [CG&E] under active criminal investigation is charged with implementing a reform program that may have been triggered by its own deliberate misconduct." Devine concluded his presentation with a plea for governmental and corporate accountability: "The public will no longer accept a program that relies on the fox to assess the strength of the henhouse." The hearings ended, but the congressional investigation continued, the outcome to be determined at a later date.[29]

In the meantime, in the summer and fall of 1982, Zimmer underwent numerous tests and inspections. During them, NRC inspectors confirmed many of Devine's concerns about the quality confirmation program. One of the big issues was that, as Devine said to the subcommittee, original quality assurance work was so inadequate that it made the reinspection work very difficult. The NRC noted, for instance, that over one hundred welders had failed to show proper qualifications, prompting commission officials to recommend that the welds done by these men (which numbered in the thousands) needed to be rechecked. The NRC also admitted that welders who no longer worked at Zimmer—some two thousand men—needed to be located and their qualifications ascertained to prove that their welding work was qualified. On top of that, documentation from earlier quality assurance work was also missing, just as Devine had said.[30]

Then, more whistleblowers came forward to Devine. In the summer of 1982 alone, he received forty-two new allegations from workers, citing ongoing issues—incorrect recordkeeping, untrained welders, welders working without any supervision, and harassment of inspectors. Whistleblowers shared that management and other workers yelled and cursed at quality assurance inspectors. Many told Devine that if a worker found something wrong, they did not have access to the highest levels of management to report violations, and even when they did report an error or issue, supervisors ignored them. According to one NRC official's best guess, this practice had led to 1,700 instances of "nonconformance" with safety standards being voided and hundreds other "missing." Whistleblowers shared that hold tags—placed on equipment or systems with problems—were frequently removed and that inspectors and other workers who brought up these issues were often dismissed. This led to many men being *Zimmerized*, the term workers used to mean being acquiescent to CG&E—to put your head down and go along.[31]

Despite the fact that the NRC was the one identifying and corroborating these issues, the agency refused to take dramatic action on Zimmer—such as stopping safety-related construction and launching a truly independent inspection of the plant. So Devine took to another strategy: he began to petition the NRC to reopen public hearings on Zimmer's operating permit. They were stalled, you'll recall, after FEMA and a licensing board failed CG&E on its accident readiness. Devine—along with the intervenors—figured that if they could get the hearings restarted, they could show an Atomic Safety and Licensing Board just how badly the quality confirmation program was going. And, their hope was that evidence would convince the board to deny CG&E a license. Ever thorough, Devine sent the NRC a 170-page summary of his investigative work and his collaboration with whistleblowers, followed by 3,000 pages of supporting documents.[32]

In response, NRC officials argued over whether to restart hearings. The Atomic Safety and Licensing Board that convened over emergency planning agreed to reconvene, but on Friday, July 30, 1982, the NRC reversed that order, refusing to restart the sessions. The commission contended that Devine's evidence was old information, nothing new to justify a hearing. With that, the licensing board authorized CG&E to load fuel and begin low-power testing at Zimmer—an order that could be reversed if the NRC decided to reopen licensing hearings. The NRC, aware of the brewing local movement against Zimmer, did say that it would permit "extensive informal public participation in monitoring corrective action at Zimmer." What that exactly meant, though, nobody knew.[33]

Representatives from the city of Cincinnati decided to hold the NRC to that promise. The city was no longer an intervenor in Zimmer's licensure progress; after much consideration, it relinquished that role in the fall of 1981 in exchange for CG&E to install additional water and air monitoring equipment at Zimmer. Given this, the Charterite-Democratic coalition in city council had to devise another way to impact Zimmer. Its Environmental Advisory Council had an idea: If the NRC refused to restart hearings, why not just convene hearings anyway? And that was exactly what the environmental council, there to advise city councilors on best policy, did. The group had twenty-three members and was then chaired by an attorney, D. David Altman, who had a decade of experience in environmental litigation and activism. Brewster Rhoads, the head of the Ohio Public Interest Campaign, and Tom Carpenter from CARE sat on the board alongside others, many of whom had scientific and public health backgrounds.[34]

Throughout the late summer and early fall of 1982, Cincinnati residents filed into city hall chambers for the Environmental Advisory Council's hearings on Zimmer. CG&E officials refused to participate, but the NRC came. So did Tom Devine and his whistleblowers. At one hearing in September, two welding supervisors agreed to come. Neither had formal training, they shared with the audience, but they were nonetheless supposed to be reviewing welding on safety systems as part of the quality confirmation program. And while they had already gone public by submitting affidavits through Devine to the NRC, the men were still terrified to be in such a public place. In fact, they only appeared on the condition that their identities would be protected since "they feared for their physical safety," as Altman told the audience. Altman, as the head of the Environmental Council, arranged for the men to sit behind a sheet-covered doorway, flanked by two police officers. Each wore a cowboy hat and used a voice-modifying machine to distort their voices. Called Mr. Q and Mr. R, the men answered a series of yes-and-no questions about the quality confirmation program at Zimmer. They said that inspectors were not thoroughly evaluating the plant to find faults. Furthermore, they had been told by CG&E and Kaiser that they should not expect to do any major repairs, even though entire sections of Zimmer needed to be totally rewelded. The audience then heard from another whistleblower: Ohio's chief boiler inspector (the state official in charge of making sure boilers and unfired pressure vessels were constructed and maintained per national engineering standards). The inspector stated bluntly and clearly that there was "no evidence" that Zimmer's reactor containment structure—the vessel where fission would occur—met engineering codes.[35]

Now, these hearings had no legal or administrative grounds to coerce the NRC to do anything, but—by being a public relations nightmare for the commission—they did help to enlarge the local circle against Zimmer. One outcome was renewed city interest in opposing Zimmer. From recommendations from its Environmental Advisory Council, city council joined Devine and other intervenors in petitioning the NRC to perform an independent audit on Zimmer.[36]

Still, the NRC refused to restart official hearings. So Devine tried another tactic: he filed a citizen's petition. For any government agency, a concerned citizen can file a citizen's petition for action on health and safety threats, and if the agency rejects or ignores it, that citizen can go to the court of appeals. On the basis that Zimmer posed a grave threat to public health and safety—and a major financial risk for CG&E, Columbus and Southern Ohio Electric Company, and Dayton Power and Light Company customers—Devine filed a citizen's petition to suspend its construction. By filing it, he released to the public all 3,170 pages of his gathered evidence that would have been entered into the licensing hearings. He then held press conferences in Cincinnati and Washington, DC, where he publicly asked the NRC to reconsider its cessation of licensing hearings.[37]

Soon after, on Thursday, November 11, 1982, a federal grand jury in Cincinnati subpoenaed Devine to testify on the matter of CG&E criminally denying Kaiser support for its quality assurance. The following day, on Friday, November 12, 1982, NRC commissioners decided to finally act. From Washington, DC, they ordered a construction shutdown of Zimmer, estimated to last for at least six months. It was the first time the NRC had paused a nuclear plant that was over 90 percent complete. Citing a "widespread breakdown" in quality assurance, the NRC ruled that all safety-related construction should stop pending an independent review of CG&E's management of Zimmer.[38]

With that victory in hand, Devine told the *Cincinnati Enquirer* that the NRC decision "restores our faith in government." And while the internal feeling among intervenors was that the plant was unsalvageable and their goal was still to stop its licensure, Devine said of the NRC ruling, "This action is the necessary first step if the plant is to be completed safely. In the long run, today's order will mean more jobs for Cincinnati construction workers." This was, again, Devine being careful to moderate any personal views he held on Zimmer's future, but his statement also reflected how he, more than anyone, worked with inspectors and construction workers at Zimmer. He was sensitive to how this community would be affected by the NRC ruling. And indeed, an immediate result of the NRC shutdown was the laying-off of fifteen hundred construction workers. And a month before that, in October 1982, CG&E management had laid off around five hundred workers after the utility said large crews were no longer needed at the plant. Prior to the shutdown, the Cincinnati Building Trades Council stated that fifteen thousand of its members were out of work because of "an already depressed construction market." The council represented sixteen craft unions within the AFL-CIO and supplied a lot of labor at Zimmer, around 10 million man-hours. Devine was well aware that Zimmer's construction stoppage would exacerbate the unemployment affecting council membership, making whistleblowers at Zimmer—already a minority—more vulnerable.[39]

Zimmer's fate continued to worsen. One month after the NRC shutdown, the congressional inquiry into the Ohio nuclear power plant confirmed Devine's charges against the NRC. On Friday, December 10, 1982, Congressman Morris Udall verified that a supervisor from the NRC's Office of Inspector and Auditor had ordered an NRC inspector to remove and hide incriminating documents against CG&E and the NRC so that Devine would not have access to the information.[40]

As that news percolated, CG&E grudgingly selected an auditor to evaluate its management of Zimmer—an auditor, mind you, that was supposed to have had no prior involvement with the plant. For this, the utility hired Bechtel Power, a subsidiary of Bechtel Corporation, a massive international corporation involved in construction, oil drilling, and nuclear power. By the early 1980s, Bechtel had completed thirty-four nuclear power plants in the US, almost half of all the nuclear stations in America, and had also begun building nuclear stations abroad, in South Korea and Taiwan. Bechtel was an obvious choice for CG&E as it had increasingly taken on the role of troubleshooter for many nuclear power plants, including remediating the damaged Three Mile Island reactor. When Bechtel became involved with Zimmer, the company had just agreed to bring three paused nuclear power plants in Washington state to completion and had replaced the original manager for two other nuclear power plants, both under construction in Texas for Houston Light and Power. It had also taken over as project manager at the Diablo Canyon nuclear plant in California and become the contractor, architect, and engineer at the one in Midland, Michigan.[41]

Devine was familiar with Bechtel. (In fact, most Americans were from the fact that the US president had filled his cabinet with Bechtel execs. Reagan's Secretary of Defense, Caspar Weinberger, was former Bechtel legal counsel; his Secretary of State, George Schultz, was a former Bechtel president; and Department of Energy deputy secretary W. Kenneth Davis was formerly Bechtel's vice president for nuclear development. The company was clearly well-connected.) But Devine specifically knew Bechtel from other whistleblowers he represented, whistleblowers at Three Mile Island, Diablo Canyon, and Midland, all of whom accused Bechtel of prioritizing speed over safety. So when CG&E hired the construction giant as its auditor, Devine told his clients that he was not surprised, but he was deeply worried.[42]

Up to this point—1982—there was no organized, cohesive movement against Zimmer. The Coalition for Affordable and Safe Energy changed that. In the summer of 1982, Tom Carpenter and Tom Devine called a meeting at Cincinnati's Xavier University to form "a broad-based, non-partisan coalition of concerned organizations within the Cincinnati region." The "non-partisan" part was critical. They invited all the people and groups already organizing against the plant, along with others they thought might be interested. In doing so, they knew they were gathering people of many backgrounds and persuasions with differing opinions on nuclear power, among other topics. To set up the coalition for success, Carpenter and Devine acknowledged in their invite, "We are at the point in this issue over Zimmer that safety is the key concern, not 'pro' or 'anti' nuclear power." The purpose of the coalition, they told people, was to oppose CG&E's "faulty workmanship, continuing lack of quality assurance and utility mismanagement at the Zimmer nuclear station. And from the meeting, CASE emerged as a coalition with one purpose: to stop the licensure of Zimmer because it was unsafe *and* too expensive."[43]

At its helm was a steering committee with rotating members. Some familiar faces occupied these leadership roles: people like Tom Carpenter and Alice Gerdeman (the nun, elementary school principal, and Zimmer Area Citizens leader). New faces joined, too, such as Kit Wood. An Ohio native, Wood had spent a summer working as a volunteer and intern for the Government Accountability Project, supporting a whistleblower client. Since she proved to be an extremely effective organizer, Devine then recruited her to help organize the coalition against Zimmer. Wood went straight to work, cold-calling people and setting up extensive phone trees to spread the word about CASE. She met with groups directly to ask them to join the coalition. Within one month of work, she convinced fifty organizations to sign on. One of her fellow activists called her "the connective tissue between people." Known for being friendly, charismatic, and able to talk to many kinds of people, she was also one of the most effectual fundraisers for the coalition.[44]

Phil Amadon was another leader, also on the younger side like Wood. In his early thirties at the time of CASE's formation, Amadon was nonetheless a seasoned activist in the labor movement. After attending college for environmental science, he became a railroad mechanic and welder, first out west in Colorado and later in Cincinnati. With his environmental science background, Amadon had long been critical of nuclear weapons, and with his labor activism, he hated how nuclear weapons production happened on the backs—the bodies—of workers. When he was working in Colorado in the 1970s, his railroad company serviced shipments to the Rocky Flats nuclear weapons production facility. One day there, a railroad car had an accident, setting off a general panic about possible radioactive contamination near the tracks. Amadon contacted activists within the local antinuclear weapons movement, knowing that his supervisors would try to quiet news of the incident. Threatened with dismissal by railroad management, Amadon stood his ground, demanding his union representative. He was not fired, but when he moved to Cincinnati for railroad welding work, his new supervisor on the repair track had heard of what Amadon did by Rocky Flats. The supervisor told Amadon, "I don't agree with how you did it, but I support what you did." When it came to Zimmer, Amadon—long before the creation of CASE—had become convinced that the plant needed to be stopped. As a proud working-class union man, he took seriously what the whistleblowers were saying. He also had a growing family, and while they did not live immediately around Zimmer, Amadon was nonetheless worried about their safety.[45]

Other leaders joined the fold. There was Dan Zavon from the Miami Valley Power Project, whom Amadon described as "the stereotypical guy in a dark room surrounded by piles of books, a pencil behind the ear, taking notes. . . . He was super focused, very smart, and very important to the effort." There was Rick Anderson, a University of Cincinnati mathematics professor, and seasoned environmental activist Paulette Meyers. Longtime Cincinnati Public Schools teacher Tim Kraus and another teacher, Marilyn Bossman, were coalition leaders as well. So was firefighter Stan Nassano. From northern Kentucky, Nassano was "a classic working-class union guy—muscular, square-jawed firefighter," as Amadon said. Bernie Reiber, a working-class woman involved in anti-CG&E rate hike organizing, and Dave Sharpenberger, another anti–rate hike organizer, also helped to lead CASE. On top of that, there was Susan Freemont from Cincinnati's Woman's City Club—described by her fellow organizers as "a very sophisticated activist, a Jane Fonda type of person"—and Vicky Mayer, another Woman's City Club member who had long been opposed to nuclear energy. (Her family had been ranchers out west and from nuclear atmospheric testing had suffered the effects of radiation poisoning.) Critically, longtime civil rights leader Marian Spencer also joined the cause.[46]

It was a whirlwind of leaders, hard to keep straight. The important part, though, was that the number and diversity of people clearly indicated that across the Cincinnati tristate region, concern over Zimmer had spread far and wide, reaching people of different incomes, professions, and even races. Some brought activist experience; many didn't. Some were younger like Amadon and Wood; others weren't. They were truly a diverse lot who largely didn't know each other before organizing together.

Many preexisting anti-Zimmer groups—like Carpenter's CARE, the Miami Valley Power Project, and Zimmer Area Citizens—became member organizations of the coalition. The Ohio Consumers' Counsel and the Citywide Coalition for Utility Reform did too, especially given that by the end of 1982, Zimmer's costs had risen to $1.7 billion. Each month the project stalled, it cost another $15 million. And with the construction shutdown, the credit rating agency Standard and Poor's downgraded CG&E's rating and added the utility and its partners in Zimmer to a credit watch list. On top of that, after the construction shutdown, the utility requested a rate hike from the Public Utilities Commission of Ohio to boost revenue by $100 million, half of which would pay for Zimmer. The utility had received an electric rate increase earlier in the year, which included Zimmer costs. So these consumer rights groups were quick to add their support to CASE.[47]

Cincinnati's Environmental Advisory Council and the city of Mentor, Kentucky, worried about emergency planning, joined too. The list kept going: local and regional chapters of the Sierra

Club and the Kentucky Conservation League also subscribed, along with the University of Cincinnati chapter of the American Association of University Professors plus student groups—including at the University of Kentucky in Lexington (about eighty miles south of Cincinnati). The local chapter of the American Federation of Government Employees, the largest federal employee union, also signed onto CASE, as did the Cincinnati chapter of the National Lawyers Guild.[48]

Indicative of Cincinnati's large Appalachian and Black populations, the local NAACP joined the coalition, as did the Urban Appalachian Council (a community organization established to help newcomers from eastern Kentucky) and the nonprofit Appalachia-Science in the Public Interest, a Kentucky group founded in 1977 to provide educational resources on healthy and sustainable living to central Appalachian communities. Across the city, neighborhood community councils voted to become a part of the coalition, and most of these were in lower-income Black and Appalachian neighborhoods.[49]

Unions representing Ohio, Kentucky, and West Virginia firefighters, first responders, and hospital workers also signed on. As frontline workers, they were particularly worried about emergency planning. Local construction trades, as well as regional coal miners' and railroad workers' unions, also became a part of the coalition. Many ordinary residents—homeowners, parents, CG&E ratepayers—paid the five-dollar fee to become members. Teachers were a large part of the coalition, underscored by the membership of the Cincinnati Federation of Teachers. Churches also enlisted themselves. Local women's groups did, too, including the Cincinnati chapter of the Women's International League for Peace and Freedom. By the summer of 1983, twenty thousand individuals and fifty organizations counted themselves as part of the Coalition for Affordable and Safe Energy.[50]

Of course, within such a sizable group, there were varying opinions about nuclear power and, among leaders, collaborative tension. Some, especially those from Carpenter's CARE, were disappointed that CASE never took a stronger stance on nuclear technology or framed the battle over Zimmer as part of a national movement against nuclear power. But others, such as Brewster Rhoads's Ohio Public Interest Campaign, organizing against CG&E's rate increases, declined to join CASE on the grounds that totalizing opposition to Zimmer was a step too far. They argued that it was better to focus on smaller goals, like ending CG&E's ability to include some of Zimmer's construction costs in its rate base. And then there was the issue among CASE leaders that its membership (and leadership) should be more exclusive: a few felt that the coalition would better serve its cause if only educated technical professionals who could fluently talk about nuclear power led it.[51]

Ultimately, CASE leaders managed to mostly bury this conflict. And many of its leaders credited its simple platform—stopping Zimmer because it was too unsafe and expensive—with CASE's broad appeal. Behind this straightforward mission was people's belief in their right to voice opposition to Zimmer and to be taken seriously by the relevant authorities—that, and their right to distrust authorities. People across the Cincinnati area were understandably desirous of more accountability with Zimmer. Those values—first expressed by David Fankhauser, then Jerry Springer, then Tom Carpenter, then Alice Gerdeman, then whistleblowers, and then utility ratepayers—had spread as Zimmer's condition worsened. By the early 1980s, they unified a lot of people, making the odd bedfellows of the coalition not that odd. CASE leaders wrote of their coalition, "All of these groups from the tri-state region charge federal, state, and local lawmakers and regulators to take the necessary political and legal steps to assure that until all safety problems are resolved that the William H. Zimmer Nuclear Power Station is not licensed and that consumers are not required to pay any further rate increases until it produces electricity." With its politically neutral and pragmatic stance, CASE managed to work alongside people and groups that were critical of Zimmer but, for various reasons, did not join CASE. To this point, Amadon and other CASE leaders collaborated with Rhoads's Ohio Public Interest Campaign, and they convinced Democratic and Republican representatives from Ohio and Kentucky to support CASE as an understandable local response to a troubled power plant. Other communities fighting troubled nuclear power plants similarly built coalitions composed of different groups with safety and consumer rights perspectives and also utilized the idea of accountability to unite people. Still, the size, diversity, and single-minded aim of CASE certainly made it stand out.[52]

While the coalition argued for CG&E to abandon Zimmer as a nuclear power facility, its leaders were aware that canceling Zimmer would not solve the matter of who would pay for the plant, nor would it resolve the area's energy needs. CASE's steering committee acknowledged this, writing, "The abandonment of Zimmer as a nuclear plant does not mean we have achieved affordable, safe energy for this area. We still have high unemployment, environmental concerns such as acid rain, escalating utility bills and uncertainty as to who will pay for the mismanagement at Zimmer." Showing that they were interested in practical solutions, the coalition advocated for a "comprehensive energy plan for this region to address issues such as energy conservation, how we can economically meet estimated future demand for electricity, the desirability of converting Zimmer to a fossil fuel facility, and the environmental impact of any action." More to the point, CASE leaders clarified what they stood for: "No money for mismanagement; the clean and efficient burning of Appalachian coal; and an energy conservation program (funded by CG&E—will give people jobs)." Of course, some organizers were disappointed that converting Zimmer to "Appalachian coal" was the best option. Adding another fossil fuel plant to an area with poor air quality was hardly ideal. (At one point, CASE debated whether to accept funds from the Bituminous Coal Institute, feeling conflicted about a direct donation from coal interests.) Nonetheless, the coalition decided that a coal plant was a more calculable risk than a dangerous and costly nuclear power plant.[53]

With such a large membership roll, the coalition's leadership aimed to use the massive body to support Devine's legal work against CG&E and the NRC. Basically, CASE was to operate as a grassroots pressure engine. The steering committee spent a lot of its energy on continually enlarging CASE. Members published op-eds and visited community groups and organizations, asking them to join and stressing that very little was required of a group for it to be a part of the coalition. All they had to do was draft a resolution of support, attend CASE's once-a-month meeting, and write to stakeholders in Zimmer—NRC officials, the Public Utilities Commission of Ohio, and elected representatives. As people joined, they were also instructed to go to the next public gathering on Zimmer. There, if people wanted to speak, the steering committee encouraged them to do so, telling them to be "their own expert" and speak from their hearts. CASE leaders—especially Kit Wood—also fundraised. Money raised went to housing and otherwise supporting Devine's whistleblowers as well as paying Devine for his legal counsel.[54]

Coalition leaders worked to their own strengths, knowing whom they would be most effective to recruit. Amadon, as a longtime labor organizer and proud union member himself, went to regional chapters (called *locals*), asking them to join. He recruited 42 United Mine Workers of America locals from Ohio and West Virginia, along with a United Steelworkers of America local in Cincinnati. Miners primarily joined the coalition because Zimmer, as a nearby nuclear power plant, took jobs from the coal industry. Members in Amadon's own railway union similarly preferred coal power for its associated railroad jobs. Still, several men in the union also lived by Zimmer, making its health impacts real, so when it came down to the vote, 80 percent of the railway workers decided to join CASE. Stan Nassano, the firefighter from Covington, Kentucky, took the lead in recruiting other unions—like the Hospital Workers Union (with locals in Ohio, Kentucky, and West Virginia) and the Covington Firefighters Union. Nassano was one of the firefighters who battled to put out the devastating Beverly Hills Supper

Club fire in Southgate, Kentucky, in 1977, which killed 165 people. He took public safety seriously, as did other firefighters and first responders. All told, Nassano and Amadon greatly enlarged local labor unions' involvement in the anti-Zimmer movement. (Of course, this was balanced against the fact that many workers who had been laid off from Zimmer after the construction shutdown were eager for work to resume. When CG&E executives held shareholder meetings in downtown Cincinnati, at company headquarters, hundreds of construction crew members picketed outside.)[55]

Marian Spencer, as a longtime civil rights activist and the president of the local NAACP, took charge of mobilizing local Black residents to engage around Zimmer. Born in Gallipolis, Ohio, in 1920, she was the granddaughter of a former slave. In 1938, she moved to Cincinnati to attend the University of Cincinnati. She met her husband, Donald, there, and the two devoted their lives to fighting for racial equality. A lifelong member of the NAACP, Spencer served multiple leadership roles within the organization, including being the first female president of the Cincinnati branch. She began her public fight for desegregation in 1952 when she and her two sons tried to visit the city's amusement park, Coney Island, and were turned away on account of their race. Joining forces with NAACP attorneys, she filed a lawsuit against the park—and won. She was also instrumental in desegregating Cincinnati's schools: for over twenty years, Spencer chaired Cincinnati's NAACP Education committee, working to achieve educational equity. In 1983, one year after CASE's formation, Cincinnati voters elected Spencer to city council on the Charter ticket. Joining the Charterite-Democratic coalition, she was the first Black woman councilor and served as vice mayor in her time in office.[56]

Spencer was well known and beloved across Cincinnati, and undoubtedly her activism primed audiences at local Black churches, where she did most of her recruiting for CASE. Until her involvement, local Black participation in the anti-Zimmer movement focused on CG&E's rate increases and poor customer service. Spencer took that momentum and expanded it. She highlighted not only CG&E's ongoing rate hikes but also that if a major accident occurred at Zimmer, it could affect the entirety of the metropolitan area, not just the immediate zone around the plant. She also stressed the NRC's lack of accountability. This was a particularly prescient issue to Black citizens: there was a long history in the US of federal officials and agencies lying to, abusing, and discriminating against African Americans. Consequently, they had good reason to be skeptical of government promises, and Spencer used that to court support for better accountability around Zimmer. At one public hearing over Zimmer in 1983, Spencer took the podium to say that she and other Black Americans were concerned over "CG&E's attitude toward public involvement in the decision-making process regarding Zimmer's future." She emphasized that CG&E and the NRC needed a watchdog to finish Zimmer properly, not quickly and cheaply. "I tell you this," she said to NRC officials. "The public can do what all the experts in the world cannot. It can reduce to their essence what the fundamental questions are."[57]

As more African Americans from around Cincinnati signed on to the anti-Zimmer movement, they underscored—and perhaps gained energy from—growing Black environmental activism across the US. While the environmental movement was initially an overwhelmingly white one, by the 1970s and 1980s, that was changing. Communities of color were realizing that environmental threats (like waste sites and hazardous production facilities) were disproportionately located near them, revealing racist decisions within the government, military, and various industries. Civil rights activists then waged local battles to relocate or remediate toxic materials. Of course, the issue with nuclear power plants, like Zimmer, was that many were purposefully located in rural, low-population zones, which, across the US (excluding the Deep South), were mostly white areas. Zimmer, then, and other nuclear power plants felt like a remote threat to many African Americans—that is, until leaders like Spencer made the issue relevant and pertinent.[58]

As that happened, Black Americans became more opposed to nuclear power as an environmental justice issue. A little over one month after Three Mile Island, upward of 125,000 people gathered in Washington, DC, to oppose nuclear power. There, African American civil rights crusader and comedian Dick Gregory addressed the crowd. Making a similar argument as Spencer, he said, "What we're doing here today is more important than dealing with racism, than dealing with sexism, than dealing with hunger, 'cause I can feel hunger! I can see war! I can feel racism! I cannot see radiation! I cannot smell radiation! I look around one day, and I am damned!" More to this point, a 1982 poll by the Institute for Policy Research at the University of Cincinnati showed that a majority of Ohioans did not want any more nuclear power plants in their state. That opposition was strongest among women, Democrats, and Black Americans.[59]

As the Coalition for Affordable and Safe Energy expanded its ranks, Tom Devine was elated. The number of Zimmer whistleblowers kept growing, and Devine finally had a unified and coordinated grassroots movement to support them, especially as CG&E moved forward with Bechtel as its auditor and construction manager. What followed the utility and the NRC in 1983 as they attempted to move forward with Zimmer—congressional charges, arbitration proceedings, class-action lawsuits, more revelations of criminal mismanagement, a shocking cost update, and a growing oppositional coalition—finally sank the nuclear power plant project.[60]

Conclusion

The End of Zimmer

Bechtel's involvement as both auditor and construction manager was short-lived. One month after the NRC ordered Zimmer's construction shutdown, Devine discovered that Bechtel had been involved in project management at Zimmer several months earlier, violating the NRC's condition that CG&E's auditor have no prior history at the plant. Consequently, Bechtel could only serve as a construction manager at Zimmer, and CG&E selected another auditor for its management review: Torrey Pines, an engineering and troubleshooting company that had just completed a review of construction safety issues at the troubled Shoreham nuclear power plant on Long Island.[1]

Over the course of 1983, as Torrey Pines prepared its audit and as Bechtel generated its final construction bid, a growing list of problems plagued the Zimmer plant: congressional charges against the NRC, shareholder lawsuits against CG&E, the PUCO denying CG&E further inclusion of Zimmer's costs in the rate base, downgraded bond ratings for Zimmer's owners, and relentless pressure against the power plant from Devine and the anti-Zimmer coalition. Altogether, these developments—along with Torrey Pines's scathing audit and Bechtel's shocking cost estimate—set the wheels in motion for the nuclear power plant's cancellation.

During the spring and summer of 1983, when Torrey Pines was reviewing Zimmer and as Bechtel put together its bid, determining what needed to be fixed at the plant and how much repairs would cost, there was no rest for CG&E and NRC officials. Zimmer was continuously in the national news for construction and financial mismanagement.

On Wednesday, April 20, 1983, Congressman Morris Udall—the head of the congressional subcommittee investigating mismanagement at Zimmer—officially charged the NRC's Office of Inspector and Auditor with purging its files of reports that showed NRC and CG&E negligence with regard to Zimmer's quality assurance. Following Udall's announcement, the NRC dismissed one of its own directors for mishandling quality assurance at Zimmer and several other nuclear plants. That man, James Cumming, was personally guilty of taking home and concealing materials Devine had requested.[2]

Around this time, the executives of Columbus and Southern Ohio Electric and Dayton Power and Light, CG&E's partners in Zimmer, began to finally lose faith in CG&E. First, the Columbus and Dayton utility leaders pressed for reduced ownership in Zimmer and urged CG&E to consider coal conversion so that money invested into the plant would not be a total waste. Then things escalated. When the three utilities formed their partnership in 1969, they had agreed to take disputes to an arbitrator instead of a courtroom—and on Thursday, January 20, 1983, Dayton Power initiated arbitration proceedings against CG&E for damages arising from CG&E's mismanagement of Zimmer.[3]

Individual stockholders also revolted against CG&E and its partners, providing fodder for CASE's platform. One Dayton Power shareholder—on behalf of themselves and 76,000 other Dayton Power and CG&E shareholders—filed a lawsuit in US District Court against CG&E for financial mismanagement, demanding the utility return to shareholders the difference between the utility's original 1969 Zimmer cost estimate and the current one, a difference of about $1.4 billion. A US district judge gave CG&E until the end of 1983 to allow the utility's litigation committee to investigate the charges. Other shareholders sent CG&E letters demanding that the utility cancel Zimmer, either abandon it or convert it to coal.[4]

CG&E leadership endeavored to placate its customers. It bought space in local newspapers, reassuring readership that it was a company that had been in the community for 150 years and that it would successfully finish Zimmer to provide "clean, safe nuclear energy." "Total quality assurance is our goal," management promised. To many across the area, such pledges were too little, too late. Tom Carpenter penned a response from the Coalition for Affordable and Safe Energy. "The Zimmer plant," he said, "is so full of bad welds, faulty piping, substandard materials and countless other physical defects that a CG&E team of hundreds of quality inspectors could not begin to identify all the problems in a year and a half's time." It was, in short, "a Pandora's box of design and construction errors."[5]

Devine continued to appeal to the NRC to reopen licensing hearings, especially because more whistleblowers came forward, even after the shutdown. A quality assurance engineer from Kaiser told Devine that in her experience at Zimmer, incorrect welding procedures had been used to build over 95 percent of the plant. When she informed CG&E managers of the problem, she was ignored—for nearly two years. Devine made her allegations public, releasing a press release that the Kaiser employee—one of the few women working at Zimmer—"was subjected to harassment, was branded a 'spy' and terminated in February [1983]." He also noted that "her conclusions were rewritten," the implication being that CG&E and Kaiser managers downplayed her concerns. The woman's affidavit joined fifty others submitted to Devine just in the summer of 1983 alone. Devine kept the leadership of the anti-Zimmer coalition informed, and they in turn alerted members to the new whistleblowers, growing the coalition's ranks with the fresh charges. Still, to Devine's great frustration, the NRC refused to restart licensing hearings, arguing that—contrary to what Devine was saying—the additional information was not new evidence, and thus fresh hearings were unjustified.[6]

On Tuesday, August 23, 1983, Torrey Pines released its 391-page review of CG&E's management of Zimmer. Its staff reviewed over 3,000 documents and interviewed around 100 people. And while the Coalition for Affordable and Safe Energy—and all the other critics of Zimmer—had doubts that the auditor would produce a critical analysis, Torrey Pines slammed CG&E, Kaiser, and the NRC for a total management breakdown in quality assurance. The auditor concluded that utility leadership had prioritized profits and costs over rigorous construction management and safety, that Kaiser was underqualified to be the general contractor, and that the NRC was far too passive in its regulatory role.[7]

CG&E, Torrey Pines wrote, had "attempted to use a project management approach that had been previously used successfully in the construction of fossil fuel plants." This included "a small, dedicated management team using relatively informal management systems and techniques" with an emphasis "on getting the plant built on schedule, at the minimum cost." "CG&E was not prepared for the complexity of the project requirements that evolved throughout the 1970's," auditing staff determined. To this end, the utility had not hired enough people for construction. Those it did hire were not qualified enough, and it had relied too much on contractors with little to no nuclear power plant experience. Utility management, Torrey Pines said, had also failed to institute procedures to ensure that installation and inspection of systems and equipment were adequate—that crucial quality assurance component to building a nuclear power station. The auditor also assessed that CG&E leaders had not created a culture friendly to critical feedback.[8]

The collective result of all these issues, Torrey Pines wrote, was that the general public now assumed the company was "guilty" until proven innocent. Nonetheless, the firm did say the project *could* be finished—but only with major changes in management and quality assurance. Torrey Pines suggested CG&E create oversight committees consisting of CG&E managers and others not previously involved at Zimmer. More to this point, the auditor recommended that management up to but not including CG&E president William H. Dickhoner should be replaced.[9]

As punitive as the review was, critics of Zimmer were disappointed that Torrey Pines did not recommend CG&E to be wholly removed from construction management alongside Bechtel, nor did it recommend a truly independent reinspection. Nonetheless, from the opposition's point of view, the audit expanded their numbers. With the Torrey Pines's report and petitioning from CASE, Cincinnati's city council decided to reactivate its intervenor rights for future licensing hearings, even though that violated its earlier agreement with CG&E. The coalition's Marian Spencer told city councilors, "It is time that the City of Cincinnati become involved in any decisions regarding the future of the Zimmer Nuclear Power Station. . . . Experts can provide technical information; representatives of the public must assess that information. We can no longer place our trust in the hands of experts any more than in the hands of CGandE. It is the experts and CGandE who are responsible for the current situation." Moved, city council not only reinstated its legal standing with the NRC but also joined Devine and the opposition in calling on CG&E and the NRC to develop an independent oversight committee on Zimmer.[10]

As the coalition grew its numbers of supporters, CG&E executives tried something new. They invited their critics to company headquarters for private talks. On Thursday, September 1, 1983, President William Dickhoner and CG&E's newest executive, retired navy vice admiral Joseph Williams Jr., met with a few CASE leaders. After the Torrey Pines review, CG&E hired additional engineers and executives for nuclear operations, one of whom was Williams. Williams was the former commander of one of the nation's first nuclear submarines. More recently, he ran a consulting firm that worked with American Electric Power's Donald C. Cook nuclear plant in Michigan. The leaders of the coalition went into the meeting suspicious that CG&E had hired Williams merely to assuage public concern, what with his experience in nuclear technology.[11]

Dickhoner and Williams greeted the group, assuring them that CG&E wanted to work with CASE—well, *some* of the coalition's members, at least. Williams explained, in veiled language, that he found certain groups within CASE, like the Zimmer Area Citizens groups, reasonable. Others in the coalition, he said, were not reasonable. These were, it turned out, all the people and groups that Devine represented. They were, Williams said, "what we in the Navy like to call rats on the ship and we can't have a situation where we have rats on the ship because they'll disrupt the functioning of the ship." Coalition leaders sat there, stunned after such a comment. Forced to regroup and respond in real time, they looked at each other and decided that Williams's aim was to divide and conquer the opposition by sectioning off its legal arm. One of the leaders there, Alice Gerdeman, put on her best principal tone of voice and told Williams that CASE would continue to include and support all of Tom Devine's clients. She and the others promptly left and were never invited back.[12]

Devine, thwarted in compelling Torrey Pines to release more stringent recommendations for Zimmer, instead turned to the press, making sure Zimmer was in local newspapers almost every day. Environmental reporters Ron Liebau at the *Cincinnati Post* and Ben Kaufman at the *Cincinnati Enquirer* spared no details on how desperate the situation was for CG&E. And Devine made sure national outlets also covered the Cincinnati power plant—to the point that Ralph Nader specifically called Zimmer the "worst managed nuclear plant in progress."[13]

Then, on Friday, October 28, 1983, Bechtel released its plan, including cost and time estimates, to finish Zimmer. By that point, CG&E and its partner utilities had already sunk $1.7 billion into Zimmer—and Bechtel estimated that the plant would require *another* $1.7 billion to finish, with a final price tag of around $3.5 billion. Furthermore, the project would take another 28 months to finish.[14]

Given that Cincinnati, Dayton, and Columbus ratepayers had already contributed $185 million to Zimmer's construction through their rate bases, utility customers were rightfully concerned that additional construction costs for Zimmer would wind up on their utility bills. And after petitioning from the Ohio Office of Consumers' Counsel and with pressure from Ohio's governor, Dick Celeste, a remarkable thing happened: in March 1983, the PUCO refused to allow CG&E to include new costs from Zimmer's construction. The commission approved only a $30.7 million annual electric revenue increase for CG&E, less than one-third of what the utility had asked for. Then, a few months later, the PUCO announced that it was launching an investigation into CG&E's financial management. This was indicative that consumer rights and calls for utility reform were even infecting state utility commissions. For the first time, commissions began to push back against power companies' frequent rate increases. Some, instead, began encouraging utilities to enact energy conservation and efficiency measures. Commissioners believed this would help utilities lower expenses and need to build fewer power plants—*and* it would help customers retain lower energy rates.[15]

Devine capitalized on customer and shareholder anger against CG&E, weaving together people's concern for safety with Zimmer's cost issues. Writing to the NRC, he explained that as shocking as Bechtel's $3.5 billion price tag was, the final number would be even higher. The majority of Bechtel's estimate, he said, was associated costs of delay, which meant that the material and labor costs to fix the plant amounted to only, in Bechtel's calculations, $253 million. To Devine, this was a woefully small amount because Bechtel was taking Zimmer "as is," meaning it did not anticipate fixing—redoing—much. Despite overwhelming evidence from both the NRC and whistleblowers that Zimmer needed substantial repairs, Bechtel took at face value what CG&E and Kaiser had already done, believing that most of it was adequate or that necessary repairs were overblown. While Bechtel's prestige might have swayed some, Devine insisted to the

NRC that Bechtel had spent more time "engineering analyses than actual repairs."[16]

The discrepancies were outstanding. The NRC had estimated that 50 percent of Zimmer welders could not be substantiated as qualified professionals; Bechtel said it was only 10 percent. The construction giant said that 98 percent of structural steel was fine, requiring very little welding review. It had no plans to consider the fresh claims over welding, as raised by whistleblowers that summer. In contrast, whistleblowers—and NRC inspectors—had collectively identified an estimated twenty-two thousand deficient welds. Similarly, Bechtel also did not plan to review the bulk of the work of concrete workers. Quality assurance inspections were also taken as is. For parts and systems that it did plan to address, Bechtel promised that any defects could be located easily and quickly by visual inspection. In a cutting summary, Devine concluded, "[Bechtel presumes] that only a small proportion of deficient hardware will be repaired; that nonexistent records will appear; that the NRC has overestimated the scope of unqualified welders by 500%; that unqualified welders, as well as unqualified inspectors who enforced inconsistent standards, would not have any impact on the hardware; and that no embedded or buried pipes or welds would be replaced."[17]

Devine appealed to the NRC to step up—to force CG&E to create a loophole-free plan, anchored by extensive documentation behind each decision, where people would be hired to do corrective work without fear of reprisal if they found faults. All decisions should be public, he said; all criteria should be public; and all exit interviews of former employees and inspectors should be public. The last requirement grew out of the fact that, just after the Bechtel report was released, CG&E and Kaiser laid off over four hundred employees, three hundred of them document-review staff. Bechtel and CG&E conducted exit interviews with a majority of them, and one-quarter of the interviewees had expressed safety-related concerns. The NRC commented at the time that it did not intend to make any exit interviews public.[18]

The NRC, attempting to contain the growing movement against Zimmer, decided to launch that "extensive informal public participation in monitoring corrective action at Zimmer" it had promised earlier. It convened a public hearing on the evening of Tuesday, November 1, 1983, at Cincinnati's downtown convention center. Over six hundred people came. The Coalition for Affordable and Safe Energy had rallied its troops, in part by disseminating news of CG&E's just-released "Course of Action." Despite its size (six hundred pages), the report said very little that was new: existing management would remain in charge, no comprehensive reinspection of Zimmer was necessary, construction would restart soon, and one citizen would be enough to represent the public on an oversight committee. As such, it wasn't hard to convince people to come to the hearing.[19]

There, NRC commissioners listened to one person after the next, each declaring the same message: cancel Zimmer. While a few individuals offered their support of the plant, urging the NRC to license Zimmer, those people were "met with boos and catcalls," as Ben Kaufman from the *Enquirer* reported. One man attended in a gas mask and a hazmat suit. Marian Spencer—just days after her successful election to council—spoke to the crowd, demanding more public oversight. "Little did we, the unsuspecting public, realize that CG&E needed a watchdog. They needed one in '75 . . . in '78. . . . and in '80. They need one now!" Coalition leader Phil Amadon—the railway welder and longtime labor rights activist—followed her speech. Walking to the front of the room, he stood at the podium, looking directly at the NRC panel in front of him. He started to speak but stopped suddenly. He then rotated the podium so that he faced the audience and his back was to the NRC. He said that he should be addressing the public, not the government. The audience burst out with applause. "We don't want an unsafe power plant in our backyards," he told people. "The ratepayers should not pay one dime for mismanagement." Fellow CASE leader Alice Gerdeman followed. She stared down the NRC officials and told them, "You are being prayed for. We are asking divine guidance for you."[20]

A little over a month later, on Friday, December 9, 1983, an Atomic Safety and Licensing Board convened a hearing at Cincinnati's downtown courthouse. Devine had not relented in his petitioning the NRC: he continued to insist his evidence was new and pertinent enough for a licensing board to consider, and because the NRC continued to refuse him, the licensing board finally met to assess the issue.

For the meeting, CASE again called in members, and they were joined by other community organizations and residents from across the Cincinnati area. People offered their support of Devine's work and echoed his criticism of CG&E and the NRC. One man said, "I've heard Mr. Devine who's done a beautiful job . . . and a number of others, who have really brought things out in the open. These Intervenors have done a massive job for the public." When it was his turn to speak, one member of CASE implored the board, "Is it like any other case you've ever heard of? Are the degree of contentions and the scope of the breakdown, the comments of Mr. Harpster [an NRC whistleblower] saying the plant and the quality assurance is out of control and that he went to Washington and had to jump up and down to get the people in Washington and the Staff of the NRC to listen—is that like anything you've heard of?" CG&E customers also made limited statements. One man vented that "97 percent of that plant is supposed to have been completed at a cost of a figure of about 1.7 billion. . . . But what I can't figure out is the prospects of finishing the other 3 percent of this plant will cost me as a ratepayer . . . 1,500 million more dollars."[21]

CG&E legal counsel disagreed with the premise of the meeting. The company's lawyer told the licensing board, "This is a classic case of people irresponsibly just coming forth, making wild accusations, hoping someday they might hit pay dirt someplace if they keep yelling loud enough." NRC officials, present at the hearing, held firm that Devine's latest information was "merely cumulative examples of deficiencies" they were already aware of—nothing to restart hearings over. Furthermore, NRC officials said the hearing outstepped the bounds of the licensing board's authority. They argued that it did not have the power to investigate whether CG&E and the NRC "have committed any material false statements." It was true that an Atomic Safety and Licensing Board *was* tasked with a limited scope of review: it was only supposed to evaluate a nuclear power plant's safety, environmental impact, and its operator's financial and accident readiness.[22]

And yet, Devine argued that the NRC and CG&E's withholding of information from the public and from the Atomic Safety and Licensing Board, by way of stifling licensing hearings, *did* impact Zimmer's safety. He explained that since the NRC would not release his latest evidence to the board, "the Atomic Safety and Licensing Boards are helpless without reliable [NRC] staff evidence." As such, Devine insisted that it *was* appropriate for the licensing board to wade into the issue.[23]

Nonetheless, the hearing ended inconclusively, although the licensing board did ask Devine to organize his evidence for it to further examine. Then, a little over a week later, on Friday, December 16, 1983, the NRC suddenly and unilaterally approved CG&E's "Course of Action" and its "Plan to Verify the Quality of Construction," which the utility had completed with Bechtel's assistance.[24]

The opposition to Zimmer was beyond frustrated. CG&E was ready, it seemed, to restart construction with no major management changes, no plans for comprehensive reinspection, and no plans for public participation other than letting one citizen sit on a board with CG&E directors. As it did at other troubled plants, Bechtel imposed a strict timeframe for the rest of the construction. It wanted the project finished in two years, and to do so, it planned to bring in its engineers from the Diablo Canyon nuclear power station in California. Diablo Canyon, Devine was quick to note, was then under serious design review for similar construction problems, as a client of his there—an engineer—had exposed.[25]

And so Devine mobilized. Using an affidavit from the Diablo Canyon whistleblower, Devine petitioned the NRC that Bechtel was a dangerous choice for construction manager at Zimmer. Under Devine's representation, the whistleblower shared that Bechtel—which was supposed to be identifying deficiencies in welding and other vital areas at Diablo Canyon—instead approved the plant's structural integrity despite major problems. When the engineer reported his findings to his managers at Bechtel, he was fired. The whistleblower further alleged that Bechtel management had also destroyed and falsified documentation of critical engineering conclusions. Devine entreated the NRC for a different outcome for Zimmer. "There is only one rational solution," he told officials, "—a comprehensive, physical reinspection of all safety-related hardware in the plant." But Devine's petition went unheeded. So when the NRC held a public hearing a few weeks later on Wednesday, January 11, 1984, giving people a chance to comment on Bechtel's plans to restart Zimmer's construction, Devine and his clients, along with the Coalition for Affordable and Safe Energy, publicly refused to attend, telling the NRC its "complete disregard for the substance and message of public input" convinced them that any future hearings were "futile gestures."[26]

In what was otherwise a very dark moment for the anti-Zimmer movement, Devine went digging and struck gold. A few days before the hearing, on Friday, January 6, 1984, an anonymous whistleblower sent Devine the deposition of a former NRC lead investigator, James McCarten, dated to the summer of 1983. McCarten had worked on the 1981 investigation into Zimmer that had resulted in the $200,000 fine for inadequate quality assurance. He resigned a few months after the investigation due to what he felt was gross misconduct by the NRC. He gave his files and a deposition to the FBI. Devine read them, realizing that they compromised NRC director James Keppler and his regional office, which oversaw Zimmer. McCarten had described the dire situation at Zimmer in 1981 as such: "We started pursuing the QC [quality control] allegation and hit pay dirt. Every single inspector that was given a QC allegation found it to be substantiated and there was a lot of real strong feelings. Everybody just came back to the trailer every night saying hey, the one you gave me, they are all screwed up in structural welding, in radiographs, in design drawings, in the electrical area. Everything we looked at and everything we have an allegation of was proven."[27]

McCarten explained in his FBI interview that Keppler did not let news of this break. Instead, the NRC director had "knowingly" provided false statements to the public during the quality confirmation program. On top of that, his staff manipulated the program's paper trail by removing certain exhibits and statements, taking away the "meat" of the report, rewriting sections to avoid blaming CG&E for mismanagement, excluding over two hundred allegations, and not reporting criminal ones to the US Attorney. The McCarten interview also revealed that Keppler knew in 1981, as CG&E launched the quality confirmation program, that the utility was falsifying records and refusing to permit the general contractor to build, inspect, and repair the plant safely. McCarten maintained that the NRC had, as a last resort, called for the 1982 construction shutdown as a way "to dodge the issue of an internal inquiry into why Region III [Keppler's branch] failed to inspect." Inspectors assumed, McCarten testified, that the plant was "so screwed up" that after the shutdown, it would never reopen.[28]

Devine captured these developments in an extensive letter to Keppler himself, telling the NRC official that such action "compromises your impartiality." Devine also sent his evidence to the Department of Justice. He announced that unless the NRC continued the construction shutdown, dismissed Bechtel as a construction manager at Zimmer, and forced CG&E to undergo a design review and verification by a truly independent auditor, he—on behalf of his clients—would sue the NRC over the issues McCarten exposed.[29]

The Coalition for Affordable and Safe Energy spread the news. Cincinnati's city council expressed its support, passing a resolution on Tuesday, January 17, 1984, that CG&E should "immediately abandon the William H. Zimmer Power Station as a nuclear plant." Even the Ohio Public Interest Campaign—which had been cautious about joining the anti-Zimmer coalition—came out against the nuclear power plant, launching a petition drive to cancel it. Brewster Rhoads, in charge of the operation, told the press, "Zimmer must be shut down immediately." A few days later, on Friday, January 20, 1984, Moody's Investor Service, the financial rating agency, lowered the ratings of CG&E, Dayton Power and Light, and Columbus and Southern Ohio Electric long-term bonds.[30]

The next day, on Saturday, January 21, 1984—a day so bitterly cold that the ground itself froze—CG&E and its partners made the painful call to terminate their nuclear power plant project. It was a decision that came after the companies had put over 15 years and $1.7 billion into Zimmer, a plant mostly finished in its construction. Given this, they resolved to do something with it, and so they decided to convert its infrastructure to a coal-powered station, which they announced the following day. At a press conference held by the presidents of the three utilities, William H. Dickhoner—CG&E's president—solemnly told the media, "It was the best decision that could have been made."[31]

It was clear that in Dickhoner's mind, CG&E was the victim in the long Zimmer story. "We were operating in a climate where we didn't know the answer," he addressed reporters. "We didn't know when the end would come. We could have gone ahead and built the finest nuclear plant ever made and still not get an operating license." To his shareholders, he wrote (a bit more objectively), "The decision was not an easy one for any of the owners. However, after obtaining a new estimate of the cost of completing the plant as a nuclear facility, and after realistically assessing the regulatory environment, we began to discuss alternatives to the nuclear option." Devine, when asked for a quote, stated, "I think the primary blame has to rest with the U.S. Nuclear Regulatory Commission. CG&E was a babe in the nuclear woods, and the NRC never taught it how to survive."[32]

Local and national press covered the issue extensively, with Zimmer making the front page of the *New York Times*. Locally, the *Cincinnati Post* wrote, "The Cincinnati Gas & Electric Co., Dayton Power & Light Co. and Columbus and Southern Ohio Electric Co. announce they will not finish the William H. Zimmer Power Station as a nuclear plant. They say they will try to convert it instead into a coal-fired power station." Howard Wilkinson from the *Cincinnati Enquirer* summed up the news: "It takes a mountain of paper to document the stormy 15-year history of the Zimmer nuclear power plant. But it took only a simple two-page statement to announce its cancellation." On the front of the *Enquirer* was an oversized image of Zimmer's cooling tower, sitting there obsolete—like the whole project. The decade-and-a-half saga of a contested nuclear power station in Cincinnati had finally ended.[33]

The same fate befell other nuclear power stations. Four days before CG&E executives decided to convert Zimmer to coal, the utility in charge of the Marble Hill nuclear power plant in Indiana canceled it. It had had major construction problems like Zimmer and was also millions of dollars overbudget. In total, Zimmer joined around 120 other abandoned nuclear power plants in the US in the 1970s and 1980s. Since several were clustered around the Great Lakes, cancellations there and across the Midwest prompted the *Chicago Tribune* to label the region "the graveyard for nuclear power plants." More cuttingly, *Forbes Magazine* writer James Cook called the entire industry "the largest managerial disaster in business history."[34]

Projects that weren't terminated subsumed massive costs. The power station in New Hampshire—begun in 1973 with an estimated cost of $1 billion—had substantial safety issues like Zimmer. A single reactor unit (at an astonishing final cost of $5.2 billion) went into operation only in 1990, after the majority-owning utility canceled the other reactor, declared bankruptcy, went through reorganization, and borrowed a significant sum of money to stay afloat. At the time, it was the fourth largest bankruptcy in US corporate history. The Shoreham nuclear power plant on

Long Island, ordered in the 1960s, was finally licensed in 1989 at a total cost of $5 billion. It was so unpopular among community members and local government officials (including the governor) that, two years later, the utility—then facing bankruptcy—agreed to shut it down in exchange for tax benefits and rate relief from New York state. It was transferred to the state for $1.[35]

Across different areas of the US, power companies canceled their nuclear power plants for the same cocktail of reasons: massive cost overruns, difficult-to-fix safety issues, seemingly unending timelines for construction and government permitting, increasingly hard-to-secure rate increases, sinking bond and credit rating issues, and public opposition. These problems, inextricably interrelated, had been driven by national developments, from the 1970s energy crises and recessions, to Three Mile Island, to a growing interest among the public in environmentalism, consumer rights, utility reform, and government accountability. And while power companies and the NRC wouldn't admit it, utility inexperience, mismanagement, and arrogance—along with NRC complacency—further played into this complex web of factors sinking nuclear power. Among all these issues, local protest shouldn't be minimized or dismissed. Communities like Cincinnati's became unwilling to pay for and tolerate such troubled power plants, and sympathetic and skilled legal counsel—Devine and others—invaluably helped these movements, believing as Devine did that without popular support, legal challenges to mismanaged nuclear power plants would go nowhere. In other words, as CASE leader Phil Amadon said, "While it was CG&E's own mismanagement and money issues that doomed Zimmer, that outcome wasn't guaranteed without popular pressure."[36]

Epilogue

Late in 1990, Zimmer finally went into operation as a 1,300-megawatt power plant burning about eight barges' worth of coal every 24 hours. To get there, the conversion process was not easy. CG&E had to procure a number of approvals, including getting an environmental impact report past the Army Corps of Engineers and Ohio EPA, which it did only in 1987. From that, the utility spent $600,000 to relocate, monitor, and preserve a large mussel population in the Ohio River, affected by the construction. The conversion also required new air pollution control equipment to handle the high-sulfur coal CG&E was using from eastern Ohio. Zimmer's solid waste—coal-burning byproducts—was taken to a specially designed clay-lined landfill about three miles east of the plant.[1]

During the conversion, CG&E and the other co-owners of Zimmer also had to sort out their legal disputes. CG&E assumed more ownership of Zimmer. Dayton Power and Light dropped its arbitration claim against CG&E. And together, they tried to recoup costs. They auctioned off surplus equipment from the plant. They sued General Electric for faulty equipment and in 1987 won a $37.3 million settlement. They also sued Sargent and Lundy, the engineering firm for Zimmer, and won $15 million. In 1985, they reached a settlement with the Public Utilities Commission of Ohio that allowed the power companies to recover half of their Zimmer costs, making them responsible for $861 million of the $1.7 billion spent on nuclear power plant construction. Ratepayers were burdened with the other half, paying it over time in their monthly utility bills. The commission capped the amount the power companies could seek from their customers after Zimmer came on line—a cap of $3.6 billion—offering ratepayers some (rather meager) protection from future rate increases.[2]

CG&E also saved money on Zimmer's conversion by using modular construction, where major components of the plant were built at other locations and shipped to the site, and the utility reused several existing facilities from the nuclear power plant. Still, the cost to convert Zimmer and initiate operations tallied to $1.3 billion. On top of that, there was another $546 million in construction interest and contingencies that accrued during the conversion process, making Zimmer's total sum—from 1969 to 1990—$3.6 billion. In the end, it cost Zimmer's owners about the same amount of money to convert the plant to coal as it would have cost to finish Zimmer as a nuclear power plant.[3]

Once operational, Duke Energy, CG&E's successor, managed the plant until 2018, and Vistra Energy took over after that. In 2022, Zimmer closed for good. During its years of operation, Zimmer was Moscow's largest employer and 90 percent of its tax revenue. For these reasons, many in the community ended up welcoming the coal station.[4]

And yet, even as a coal power plant, Zimmer wasn't wholly popular. It continued to elicit similar reactions as when it was a nuclear power station in the making: protest by residents and workers over environmental and safety issues and protest by ratepayers and their allies who believed CG&E and the other utilities were unfairly shouldering Zimmer's costs on customers. Still, crowds at public hearings in the mid- and late 1980s, and public outcry thereafter, were significantly smaller in size—a few hundred individuals—than the peak of people showing up to hearings during Zimmer's nuclear years. As a nuclear power station, the gross number of safety-related construction errors—and the massive cost overruns, driven by mismanagement and hands-off regulation—loomed over the Cincinnati area as an acute, pressing threat, gradually uniting thousands. Zimmer as a coal power plant had real health risks, too, and, of course, continued to rack up major costs, but these issues didn't register in the same dire way to most people. The sense of crisis, for many, had passed.[5]

Ben Kaufman, the longtime *Cincinnati Enquirer* reporter who covered Zimmer for all those years, told me, "Zimmer took up too much of my life." That's certainly understandable given it took twenty-one years just to get the plant licensed to operate. One silver lining of the whole saga, though, is how it affected many of the people involved. Tom Devine, for instance, went on to serve as the Government Accountability Project's legal director, establish national and international whistleblower protection legislation, and represent over seven thousand whistleblowers. Tom Carpenter became an attorney: he managed GAP's nuclear oversight campaign from 1985 to 2007 and then ran the Hanford Challenge, the nonprofit watchdog and advocacy organization that has assisted in the cleanup of the plutonium-producing Hanford Nuclear Site in Washington state. David Fankhauser joined—and was instrumental in—the citizen effort to close and remediate the Fernald Feed Materials Production Center near Cincinnati, which produced—and was leaking—high-purity uranium metals.

Alice Gerdeman became the director of the Intercommunity Justice and Peace Center in 1992, a Cincinnati nonprofit that advocates for pro-immigration reform, international peace, economic justice, and an end to the death penalty. Brewster Rhoads served as the executive director of Green Umbrella, a nonprofit alliance that coordinates numerous environmental and outdoors organizations across the Cincinnati area. His colleague Roxanne Qualls became the Ohio Public Interest Campaign's local director and then was voted into office as Cincinnati's mayor from 1993 to 1999. Civil rights legend Marian Spencer served her two-year term on Cincinnati's city council from 1983 to 1984, the first Black woman to do so, and then represented the Ohio Democratic Party at its 1984 and 1988 national conventions. In these ways, among many others, the botched nuclear power station served as a springboard for sustained future activism and advocacy work, even as anti-Zimmer groups disbanded in 1984.

And what about the rest of us? What does the story of Zimmer leave us with?

David Fankhauser and others who participated in early public hearings teach us that all movements start somewhere, however small, and that early lone voices aren't "crazy," as CG&E and others said, but instead discerning and brave. Jerry Springer and other city councilors in the 1970s and 1980s remind us that many kinds of people can be activists, including our government representatives. The hundreds of people who attended Zimmer's hearings after Three Mile Island impart to us the power of a crowd—and the unique power of a crisis to drive us to action. Tom Carpenter, Polly Brokaw, and their fellow activists in Citizens against a Radioactive Environment highlight how we need organizations with big ambitious goals and messages, contextualizing our world for us and centering local campaigns within bigger national fights. On the flip side, CARE's growing numbers and strategic evolution over the years show us how pragmatic and powerful locally focused activism can be.

The Coalition for Affordable and Safe Energy—led by Kit Wood, Phil Amadon, Marian Spencer, Alice Gerdeman, and many others—so highlights this. That broad-based group stands as a reminder of how people within a community often have more in common than is readily apparent. Turns out, when we agree on certain fairly-easy-to-agree-on values—like public safety and corporate responsibility—we can get a lot accomplished, even if only for a short and temporary period of time. Tom Devine and his whistleblowers similarly convey that people's interest in accountable leadership—in government and politics, in science and medicine, in many areas of life—is a project that many of us can get behind, despite other differences.

From Brewster Rhoads, Roxanne Qualls, and all the others fighting for fair utility rates, we recall the power of consumer rights.

Their activism around energy conservation is prescient in our era of climate change when so many of us are thinking about where our energy comes from, how much we (really) need, how much it should cost us, how much it should cost the earth—and how these variables are often, frustratingly, at odds with one another.

Altogether, the anti-Zimmer movement left ten thousand pages of transcript from licensing hearings. People took advantage of new administrative procedures, like public hearings, to insert themselves into what otherwise were highly bureaucratic and closed-off decision-making processes, making them more democratic in the process. We benefit from their efforts today: in every neighborhood council meeting, every public government hearing, every legal challenge where a community seeks redress against a corporation or government agency.

And what about the executives behind Cincinnati Gas and Electric? And the officials in the Atomic Energy Commission and the Nuclear Regulatory Commission? Those men, who operated for so long without public oversight, with deep confidence in their right to do so, were ultimately brought to heel. When federal regulators and CG&E's directors refused to *really* consider changing values, when they failed to approach a new technology with appropriate caution, they found themselves face to face with a twenty-thousand-people-strong protest movement, a mismanaged and poorly built power plant, and massive debt. So these leaders, in their pride and stubbornness, are a reminder that times change and those in positions of authority must be aware of and amenable to societal shifts.

More sympathetically to CG&E leaders, their struggles are a tale as old as time: that businesses (especially ones that involve construction) are difficult to operate successfully and even more so during times of economic decline. And when we think about CG&E executives and the officials in charge of the AEC and NRC during Zimmer's licensure process, we can appreciate that these individuals *truly* thought of themselves as leaders who knew best and were doing their best. They teach us the importance of perspective when we tell stories, when we recount the past.

That Zimmer occurred alongside many other fraught nuclear power plants in the 1970s and 1980s reminds us of a simple truth: that knowing our past is important, especially because nuclear power remains a divisive—and increasingly talked about—issue. With passive safety improvements in nuclear reactors that do not require human intervention; with nuclear power's ability to provide continuous energy in large, centralized stations; and with our need to move away from fossil fuels, nuclear energy's benefits seem more favorable than they have been for many years. On top of that, there's the reality that many of the communities with a nuclear power station—rural, otherwise depressed areas—greatly rely on the jobs and tax income from their power plants. But with memories of the Chernobyl and Fukushima nuclear power plant disasters, many people in the United States remain skeptical. There's also the reality that we still do not have a long-term solution for the safe disposal of waste from nuclear power plants. Nuclear energy's foes call attention to these issues and others facing the industry. For the nuclear power stations still operable, their old age makes them prone to malfunction, and for the construction of new nuclear plants, the same issues that held up Zimmer—construction delays, high costs, and quality assurance issues—remain a sticky problem. Whistleblowers continue to emerge, and Tom Devine's firm continues to represent them. All of this is to say that when we think about nuclear power today, we should understand its history not that long ago to ensure a better outcome today—for people's safety, for utility customers' bills, and for the financial health of power companies.

Finally, Zimmer is about so much more than just nuclear power. It is foremost a story of a community that challenged a "higher-up" decision that was in many ways out of its control—a decision that implicated people's health and safety and their personal finances. In this way, Zimmer has powerful echoes with many other places—both places like Cincinnati, where activists preempted a public health disaster, and places where the damage had already been done. Places like Love Canal, New York, in the

1970s, when parents fought a chemical company's dump in their neighborhood by demanding remediation, relocation, and compensation from the state and federal government; like Woburn, Massachusetts, in the same years, when community members sought restitution against two corporations that had deposited industrial solvents into their drinking water, contributing to an epidemic of childhood leukemia; like Warren County, North Carolina, in the 1980s, when residents organized against the establishment of a hazardous waste facility in their county; like Hinkley, California, in the 1990s, when residents sued their local utility for contaminating their groundwater with a cancer-causing heavy metal (a case made famous in the 2000 film *Erin Brockovich*); like Flint, Michigan, in the 2010s, when—following the news that lead and other health hazards were in people's drinking water—citizens sued city and state officials for lead service line replacements and demanded the delivery of bottled water and other safety measures.

Across these different times and places, ordinary people took a stand, attempting to achieve a positive outcome for their community. Of course, it wasn't perfect. Like with Zimmer, many of these movements were fragmented: some people didn't get along; some activists didn't welcome others. And in terms of the outcomes, look at Zimmer: the movement against it wound up with a coal power plant. In that, we are reminded that activism, even with the best intentions, rarely has a perfect ending. Instead, hard trade-offs are the norm. Nonetheless, Zimmer and these other community struggles teach us, in the words of David Fankhauser, that "sometimes communities can successfully resist unhealthful corporations that have so much power."[6]

Acknowledgments

In the years I've been working on this book, I have received considerable help from many people, including numerous kind and helpful archivists and librarians. Special thanks to Kevin Grace at the University of Cincinnati's Archives and Rare Books Library and to Mickey deVise and others at the Cincinnati Museum Center's Cincinnati History Library and Archives. Thank you also to Anne Perrera Goel, the technical librarian at the Nuclear Regulatory Commission's Public Document Room in Rockville, Maryland, and to archivist Akeem Flavors at the Duke Energy Corporate Library and Archives.

Two activists in the anti-Zimmer battle—Phil Amadon and David Fankhauser—kindly let me borrow their personal collections of records. They also let me interview them (multiple times). Thank you both. Other activists involved in Zimmer also generously lent their time to speak with me: thank you to Tom Carpenter, Tom Devine, Alice Gerdeman, Stan Hedeen, Brewster Rhoads, and Roxanne Qualls. A huge thank-you to Ben Kaufman, whose journalism helped me understand Zimmer; moreover, Ben kindly read and edited a draft of this manuscript, offering invaluable—and endlessly witty—advice.

In this regard, I owe another huge thank-you to Brittany Clair for not only reading this story and offering vital feedback but also being a wonderful friend and colleague over many years.

David Stradling also kindly read, edited, and commented on this book at its various stages, and for that I am very grateful. Thank you also to Maura O'Connor and Stephen Porter, who in one way or another shaped this project. To all three of you, your mentorship and friendship mean so much to me.

This book owes its existence to the University Press of Kentucky. I am especially grateful to my acquisitions editors, Patrick O'Dowd and Natalie O'Neal Clausen, who have been so encouraging and helpful in bringing the project to life. Thank you also to Dane Ritter for guiding this project along.

Most of all, thank you to my family, especially my parents and my husband, John, who endlessly encouraged this project. And last—to Jack. This book is for you.

Notes

1. "More Economical Power"

1. "The Weather," *Cincinnati Post*, September 12, 1969.

2. "Power from Atom," *Cincinnati Enquirer*, September 13, 1969.

3. Bob Womank, "UC Professor Takes Issue with Proposed Nuclear Ban," *Cincinnati Post*, February 20, 1976.

4. Duke Energy, "A Look Back," accessed March 2, 2023, https://www.duke-energy.com/our-company/cincinnati/a-look-back; "History of Cincinnati Gas & Electric Company," accessed March 2, 2023, https://www.referenceforbusiness.com/history2/36/CINCINNATI-GAS-ELECTRIC-COMPANY.html.

5. Walter R. Keagy, *The CG&E Story* (Cincinnati: Cincinnati Gas and Electric Company, 1959); Walter C. Beckjord, *"The Queen City of the West"—During 110 Years! A Century and 10 Years of Service by the Cincinnati Gas & Electric Company (1841–1951)* (Cincinnati: Public Library of Cincinnati and Hamilton County), 10–15 (hereafter PLCHC).

6. Keagy, *The CG&E Story*; Beckjord, *"The Queen City of the West,"* 18–27; Cincinnati Gas & Electric Company (hereafter CG&E), *The Power to Serve: CG&E's Commitment to the Community* (Cincinnati: Cincinnati Gas and Electric Company, 1987), PLCHC.

7. Richard F. Hirsh, *Power Loss: The Origins of Deregulation and Restructuring the American Electric Utility System* (Cambridge, MA: The MIT Press, 1999), 26–27; Andrew Needham, *Power Lines: Phoenix and the Making of the Modern Southwest* (Princeton, NJ: Princeton University Press, 2014), 76–81.

8. Hirsh, *Power Loss*, 46–50.

9. Needham, *Power Lines*, 9–11; Hirsh, *Power Loss*, 2, 23–24, 45.

10. Thomas Borstelmann, *The 1970s: A New Global History from Civil Rights to Economic Inequality* (Princeton, NJ: Princeton University Press, 2012), 53–54.

11. CG&E, "Annual Report," 1960–1969, Duke Energy Corporate Archive, Charlotte, NC (hereafter DE).

12. Ibid.

13. Robert Fresco, "Power without Pollution—It Doesn't Come Easily," *Cincinnati Enquirer*, October 17, 1971.

14. Ibid.; Douglas Starr, "'Power Alley' Generates Problems," *Cincinnati Post*, May 30, 1980.

15. "Services Set Friday for 'Great Cincinnatian,'" *Cincinnati Enquirer*, April 3, 1973.

16. Author's interview with Stan Hedeen, October 23, 2019; CG&E, "Annual Report," 1969, DE; Beckjord, "*The Queen City of the West*," 22.

17. David Stradling, *Cincinnati: From River City to Highway Metropolis* (Charleston, SC: Arcadia Publishing, 2003), 125; "History of Cincinnati Gas & Electric Company," accessed March 2, 2023, https://www.referenceforbusiness.com/history2/36/CINCINNATI-GAS-ELECTRIC-COMPANY.html.

18. CG&E, "Annual Report," 1960–1969, 1970, DE; Beckjord, "*The Queen City of the West*," 30.

19. Hirsh, *Power Loss*, 55; quoted in Needham, *Power Lines*, 79.

20. CG&E, "Annual Report," 1965–1969, DE; "CG&E Joins 8-State Network to Prevent Mass Blackout," *Cincinnati Post*, February 6, 1967; CG&E, "Annual Report," 1969, DE; Robert D. Lifset, *Power on the Hudson: Storm King and the Emergence of Modern Environmentalism* (Pittsburgh: University of Pittsburgh Press, 2014), 18–19; Needham, *Power Lines*, 126–128.

21. The Tennessee Valley Authority, a federally owned utility, also chose to build nuclear power plants in coal country; see Thomas R. Wellock, *Critical Masses: Opposition to Nuclear Power in California, 1958–1978* (Madison: The University of Wisconsin Press, 1998), 72. CG&E, "Annual Report," 1971, DE; Robert Fresco, "Power without Pollution—It Doesn't Come Easily," *Cincinnati Enquirer*, October 17, 1971; Needham, *Power Lines*, 76–79, 173–175.

22. Megan Lenore Chew, "Power from the Valley: Nuclear and Coal in the Postwar U.S." (PhD diss., Ohio State University, 2014), 59–64; J. Samuel Walker, *Three Mile Island: A Nuclear Crisis in Historical Perspective* (Berkeley: University of California Press, 2004), 4.

23. Walker, *Three Mile Island*, 29–30. For more on early government promotion and involvement in nuclear power, see also Brian Balogh, *Chain Reaction: Expert Debate and Public Participation in American Commercial Nuclear Power, 1945–1975* (New York: Cambridge University Press, 1991); and J. Samuel Walker, *Containing the Atom: Nuclear Regulation in a Changing Environment, 1963–1971* (Berkeley: University of California Press, 1992).

24. Walker, *Three Mile Island*, 5–7.

25. Walker, *Three Mile Island*, 6–7; Joan B. Aron, *Licensed to Kill?: The Nuclear Regulatory Commission and the Shoreham Power Plant* (Pittsburgh: University of Pittsburgh Press, 1997), 13–14; Gene Smith, "A Building Boom for Nuclear Power Plants," *New York Times*, January 14, 1973.

26. CG&E, "Annual Report," 1969, DE; "Atomic Power for CG&E," *Sycamore Messenger*, September 25, 1969; William Styles, "Electrical Power Situation in City Tight, Not Critical," *Cincinnati Post*, June 16, 1970; Davilynn Furlow, "The Ohio Valley around Us Is on a Power Trip," *Cincinnati Post*, December 7, 1973; "Nuclear Power Plant to Be First of Four in Clermont," *Cincinnati Post*, September 13, 1969; CG&E, "Summary of Application for Construction Permit, Wm. H. Zimmer Nuclear Power Station," 2-1, PLCHC.

27. CG&E, "Summary of Application," 3-1, PLCHC; "Zimmer Plant Is on the Grow," *Cincinnati Post*, July 18, 1973; "Zimmer Power Plant Nearing Finish," *Cincinnati Post*, August 29, 1977; CG&E, "Annual Report," 1972; Nuclear Regulatory Commission (hereafter NRC), "Draft Environmental Statement Related to the Operation of the Wm. H. Zimmer Nuclear Power Station," October 1976, personal collection of David Fankhauser, Cincinnati, OH (hereafter DF); CG&E, "Annual Report," 1969, DE.

28. CG&E, "Summary of Application," 1-1, PLCHC.

29. Ibid., 2-4, PLCHC.

30. "Vas You Efer in Moscow (Ohio, That Is)?" *Cincinnati Enquirer*, December 3, 1967; Keith BieryGolick, "The Death of a Power Plant, Maybe a Village," *Cincinnati Enquirer*, July 10, 2022.

31. CG&E, "Summary of Application," 4-1–4-11, PLCHC; NRC, "Draft Environmental Statement," DF.

32. CG&E, "Summary of Application," 4-1, PLCHC; Atomic Energy Commission (hereafter AEC), "Environmental Statement Related to Construction of the Wm. H. Zimmer Nuclear Power Station," September 1972, DF.

33. George T. Mazuzan, "'Very Risky Business': A Power Reactor for New York City," *Technology and Culture* 27, no. 2 (April 1986): 262–284.

34. Scott Aiken, "CG&E Shareholders Back A-Plant," *Cincinnati Enquirer*, April 26, 1979.

35. "Nuclear Power Plant to Be First of Four in Clermont," *Cincinnati Post*, September 13, 1969; Douglas Starr, "'Power Alley' Generates Problems," *Cincinnati Post*, May 30, 1980.

36. "'Sleepy' Moscow Wide Awake on Power Plant," *Cincinnati Enquirer*, August 22, 1971.

37. CG&E, "Annual Report," 1969, DE.

38. CG&E, "Annual Report," 1971, 1969–1980, DE.

39. CG&E, "Annual Report," 1977, 1978, DE.

40. Thomas R. Wellock, *Safe Enough: A History of Nuclear Power and Accident Risk* (Oakland: University of California Press, 2021), 45.

41. "Vas You Efer in Moscow (Ohio, That Is)?" *Cincinnati Enquirer*, December 3, 1967; Keith BieryGolick, "The Death of a Power Plant, Maybe a Village," *Cincinnati Enquirer*, July 10, 2022.

42. AEC, *Hearing in the Matter of Cincinnati Gas & Electric Company*, June 1, 1972, Nuclear Regulatory Commission Records, Public Document

Room, US Nuclear Regulatory Commission Archive, Rockville, MD (hereafter NRC); CG&E, "Summary of Application," PLCHC; author's interview with David Fankhauser, October 31, 2018; AEC, "Environmental Statement Related to Construction," DF.

43. Interview with Phil Amadon, June 25, 2014; "'Sleepy' Moscow Wide Awake on Power Plant," *Cincinnati Enquirer*, August 22, 1971; for more on postwar communities accepting nuclear facilities near them for their economic benefits, see Kate Brown, *Plutopia: Nuclear Families, Atomic Cities, and the Great Soviet and American Plutonium Disasters* (New York: Oxford University Press, 2013).

44. "'Sleepy' Moscow Wide Awake on Power Plant," *Cincinnati Enquirer*, August 22, 1971; see Balogh, *Chain Reaction.*

45. AEC, *Hearing in the Matter of Cincinnati Gas & Electric Company*, June 1, 1972, NRC; "Springer Nuclear Views Protested," *Cincinnati Enquirer*, February 20, 1976.

46. AEC, *Hearing in the Matter of Cincinnati Gas & Electric Company*, June 1, 1972, NRC.

47. "'Sleepy' Moscow Wide Awake on Power Plant," *Cincinnati Enquirer*, August 22, 1971.

2. "Every Single Dose of Radiation Carries a Finite Risk"

1. AEC, *Prehearing in the Matter of Cincinnati Gas & Electric Company*, May 12, 1972, NRC.

2. Elysse Winget, "UC Professor Honored for Contributions to Civil Rights Movement," *News Record*, October 19, 2014; Lucy May, "How a White Cincinnati Freedom Rider Helped, Learned from Black Civil Rights Activists in the '60s," *WCPO News*, May 20, 2019, https://www.wcpo.com/news/our-community/how-a-white-cincinnati-freedom-rider-helped-learned-from-black-civil-rights-activists-in-the-60s.

3. AEC, *Prehearing in the Matter of Cincinnati Gas & Electric Company*, May 12, 1972; AEC, *Hearing in the Matter of Cincinnati Gas & Electric Company*, June 1, 1972, NRC.

4. AEC, *Prehearing in the Matter of Cincinnati Gas & Electric Company*, May 12, 1972, NRC; David Fankhauser, "Dear President Carter," April 26, 1977, DF.

5. AEC, *Prehearing in the Matter of Cincinnati Gas & Electric Company*, May 12, 1972, NRC.

6. CG&E, "Annual Report," 1972, DE.

7. Ben L. Kaufman, "CG&E Has a Lot of Dollars at Stake," *Cincinnati Enquirer*, July 9, 1978.

8. "How Safe Will Nuclear Waste Be at Zimmer?" *Cincinnati Enquirer*, July 2, 1978.

9. Ben L. Kaufman, "Is Maxey Flats Nuclear Burial Ground Safe," *Cincinnati Enquirer*, March 28, 1976; James Bruggers, "Maxey Flats Nuke Dump: What Were They Thinking," *Courier Journal* (Louisville), June 2, 2015; J. Samuel Walker, *The Road to Yucca Mountain: The Development of Radioactive Waste Policy in the United States* (Berkeley: University of California Press, 2009), 126; Caroline Peyton, "Kentucky's 'Atomic Graveyard': Maxey Flats and Environmental Inequality in Rural America," *Register of the Kentucky Historical Society* 115, no. 2 (Spring 2017): 223–263, 237.

10. AEC, *Hearing in the Matter of Cincinnati Gas & Electric Company*, June 1, 1972, NRC.

11. AEC, *Prehearing in the Matter of Cincinnati Gas & Electric Company*, May 12, 1972, NRC.

12. AEC, "Environmental Statement Related to Construction of the Wm. H. Zimmer Nuclear Power Station," September 1972, DF.

13. Ibid.; AEC, "Safety Evaluation by the Division of Reactor Licensing . . . in the Matter of the Cincinnati Gas and Electric Company William H. Zimmer Nuclear Power Station Unit No. 1," February 18, 1972, DF.

14. AEC, *Prehearing in the Matter of Cincinnati Gas & Electric Company*, May 12, 1972, NRC.

15. See Nancy Langston, *Toxic Bodies: Endocrine Disruptors and the Legacy of DES* (New Haven, CT: Yale University Press, 2010); Leslie Reagan, *Dangerous Pregnancies: Mothers, Disabilities, and Abortion in Modern America* (Berkeley: University of California Press, 2010); Janet Golden, *Message in a Bottle: The Making of Fetal Alcohol Syndrome* (Cambridge, MA: Harvard University Press, 2005).

16. AEC, *Prehearing in the Matter of Cincinnati Gas & Electric Company*, May 12, 1972, NRC.

17. James H. and Nancy Fichter in 1930, 1940, and 1950 Federal Censuses; James H. and Nancy Fichter Marriage Certificate in Ohio, US Country Marriage Records, 1774–1993, ancestry.com; "Jehovah's Witnesses Sent to Federal Prison," *Cincinnati Enquirer*, November 4, 1943; AEC, *Prehearing in the Matter of Cincinnati Gas & Electric Company*, May 12, 1972; AEC, *Hearing in the Matter of Cincinnati Gas & Electric Company*, June 1, 1972, NRC.

18. AEC, *Prehearing in the Matter of Cincinnati Gas & Electric Company*, May 12, 1972; AEC, *Hearing in the Matter of Cincinnati Gas & Electric Company*, June 1, 1972, NRC.

19. Author's interview with David Fankhauser, August 5, 2017.

20. See Aron, *Licensed to Kill*; Henry F. Bedford, *Seabrook Station: Citizen Politics and Nuclear Power* (Amherst: University of Massachusetts Press, 1990), 27; and John Willis, *Conservation Fallout: Nuclear Protest at Diablo Canyon* (Reno: University of Nevada Press, 2006); Hirsh, *Power Loss*, 65–66.

21. AEC, *Prehearing in the Matter of Cincinnati Gas & Electric Company*, May 12, 1972; AEC, *Hearing in the Matter of Cincinnati Gas & Electric Company*, June 1, 1972, NRC.

22. For more on Storm King Mountain, see Lifset, *Power on the Hudson*. For more examples of intervention in hearings for contested nuclear power plants, see Bedford, *Seabrook Station*; Willis, *Conservation Fallout*, specifically chap. 3; and Chew, "Power from the Valley," 130–132.

23. Troy B. Conner Jr., "Applicants' Combined Answer to Amended Petitions to Intervene of Dr. David Fankhauser, Mr. & Mrs. J. Hal Fichter and the Ohio Valley Group Concerned about Nuclear Pollution," May 16, 1972; Troy B. Connor and Mark J. Wetterhahn, "Applicants' Consolidated Reply to Petitions for Leave to Intervene of Miami Valley Power Project, David B. Fankhauser, Mari B. Leigh, and the City of Cincinnati," November 7, 1975, DF; Jo-Ann Albers, "Would-Be Intervenors Miffed over A-Plant," *Cincinnati Enquirer*, May 28, 1972.

24. AEC, *Hearing in the Matter of Cincinnati Gas & Electric Company*, June 1, 1972; AEC, *Prehearing in the Matter of Cincinnati Gas & Electric Company*, May 12, 1972, NRC.

25. AEC, *Prehearing in the Matter of Cincinnati Gas & Electric Company*, May 12, 1972, NRC; see James Longhurst, *Citizen Environmentalists* (Medford, MA: Tufts University Press, 2010), on this "active citizenship."

26. "John Basil Farmakides Obituary," May 11, 2019, https://obits.cleveland.com/us/obituaries/cleveland/name/john-farmakides-obituary?id=12940984. For more on the lack of trust—and the cultural gap—between government officials and the public when it came to nuclear power, see Aron, *Licensed to Kill*, focused on the Shoreham nuclear power plant.

27. AEC, *Hearing in the Matter of Cincinnati Gas & Electric Company*, June 1, 1972, NRC.

28. Ibid.; "AEC Hearing," *Cincinnati Post*, June 2, 1972; Richard Gibeau, "Power Plant Hearing," *Cincinnati Post*, September 20, 1972; author's interview with Stan Hedeen, October 23, 2019.

29. For more on the rise of activist-scientists, see Longhurst, *Citizen Environmentalists*, and for more on internal government dissent on nuclear power, see Balogh, *Chain Reaction*; and Wellock, *Safe Enough*.

30. AEC, *Prehearing in the Matter of Cincinnati Gas & Electric Company*, May 12, 1972, NRC.

31. Wellock, *Safe Enough*, 1–10; AEC, *Hearing in the Matter of Cincinnati Gas & Electric Company*, June 1, 1972, NRC.

32. "AEC Hearing," *Cincinnati Post*, June 2, 1972.

33. Ibid.; AEC, *Hearing in the Matter of Cincinnati Gas & Electric Company*, June 1, 1972, NRC.

34. Wellock, *Safe Enough*, 68, 11–81.

35. Richard Gibeau, "AEC Hearing," *Cincinnati Post*, June 2, 1972; AEC, *Hearing in the Matter of Cincinnati Gas & Electric Company*, June 1, 1972, NRC.

36. "AEC Hearing," *Cincinnati Post*, June 2, 1972; AEC, *Prehearing in the Matter of Cincinnati Gas & Electric Company*, May 12, 1972, NRC; author's interview with David Fankhauser, August 5, 2017; David Fankhauser, "Forum: Anticipating Zimmer's Ill Wind," *Cincinnati Post*, October 23, 1975.

37. Brown, *Plutopia*, 52, 332; see also Balogh, *Chain Reaction*; and Wellock, *Safe Enough*.

38. J. Samuel Walker, *Permissible Dose: A History of Radiation Protection in the Twentieth Century* (Los Angeles: University of California Press, 2000), 29–44, 54–66, 99–102; Brown, *Plutopia*, 308, 276.

39. AEC, *Prehearing in the Matter of Cincinnati Gas & Electric Company*, May 12, 1972; AEC, *Hearing in the Matter of Cincinnati Gas & Electric Company*, June 1, 1972, NRC.

40. Walker, *Three Mile Island*, 29–33, 40–42; Hirsh, *Power Loss*, 171.

41. Ben L. Kaufman, "Four Win Hearing to Protest Power Station," *Cincinnati Enquirer*, March 30, 1976; Samuel W. Jensch, "Order Granting Motion for Reconsideration of Form of Intervenor Contentions," April 13, 1976, DF.

42. Walker, *Permissible Dose*, 67–90; AEC, *Hearing in the Matter of Cincinnati Gas & Electric Company*, June 1, 1972; NRC, *Prehearing in the Matter of Cincinnati Gas & Electric Company*, January 23, 1976, NRC.

43. "Samuel W. Jensch," in US, Social Security Applications and Claims Index, 1936–2007; "Samuel W. Jensch," in US, School Yearbooks, 1900–2016, ancestry.com; Bratley Family Funeral Homes, "Beverly M. Jensch Obituary," October 11, 2017, https://www.bratleyfamilyfuneralhomes.com/obituary/4406645.

44. NRC, *Prehearing in the Matter of Cincinnati Gas & Electric Company*, January 23, 1976, NRC.

45. Ibid.; Stephanie Hemphill, "The Legacy of Reserve Mining Case," Minnesota Public Radio, October 29, 2003.

46. NRC, *Prehearing in the Matter of Cincinnati Gas & Electric Company*, January 23, 1976, NRC.

47. Ibid.

48. Reports from DF.

49. Thomas A. Luebbers and Peter Heile, "Intervenor City of Cincinnati's First Set of Interrogatories to Applicant," December 9, 1977, DF.

3. "The Public Has to Make the Decisions"

1. AEC, *Hearing in the Matter of Cincinnati Gas & Electric Company*, June 1, 1972, NRC; "A-Plant Site Grading 'Outdated,'" *Cincinnati Enquirer*, June 2, 1972; Richard Gibeau, "AEC Hearing," *Cincinnati Post*, June 2, 1972.

2. Howard Wilkinson, "The World Knows One Jerry Springer; We Know Another," *WVXU News*, July 13, 2018, https://www.wvxu.org/politics/2018-07-13/the-world-knows-one-jerry-springer-we-know-another.

3. "Springer for Council . . . WHY," June 1971, Springer Papers, US-71-07, Folder 1; Gerald Springer, "Announcement of Candidacy—Jerry Springer," Springer Papers, US-87-6, Folder 19; Gerald Springer, "Saylor Park," Springer Papers, US-87-6, Folder 19; "Jerry Springer's First Term in City Council," Springer Papers, US-87-6, Folder 19; "City Hall in Mt. Washington?" Springer Papers, US-71-07, Folder 6; "Vote Yourself a Congressman," Springer Papers, US-71-07, Folder 11; "Environmental Management," Springer Papers, US-71-07, Folder 11; "The State of America," Springer Papers, US-71-07, Folder 11; "A Program for the 1970's," Springer Papers, US-71-07, Folder 11, Blegen Library, University of Cincinnati Archives and Rare Books Library, Cincinnati, OH (hereafter BL); Tom Hayes, "Jerry Springer: The Cincinnati Kid," *Clifton Magazine*, Winter 1974–1975; David S. Mann, "Scene: A Smoke-Filled Room, Action: The Election of Mayor Sterne," *Cincinnati Magazine*, March 1976.

4. Stradling, *Cincinnati*, 77–81, 96; Zane L. Miller and Bruce Tucker, "The New Urban Politics: Planning and Development in Cincinnati, 1954–1988," in *Snowbelt Cities: Metropolitan Politics in the Northeast and Midwest since World War II*, ed. Richard M. Bernard (Bloomington, Indiana University Press, 1990), 92.

5. Miller and Tucker, "The New Urban Politics," 94, 96.

6. Ibid., 93–95, 98.

7. Ibid., 97; Dan Horn, "Voter Turnout Tuesday in Cincinnati and Hamilton County Is the Lowest Since at Least Mid-1970s," *Cincinnati Enquirer*, November 2, 2021.

8. University of Cincinnati Libraries, "Theodore M. Berry Papers—Timeline," accessed March 30, 2023, https://libapps.libraries.uc.edu/exhibits/berry/timeline/; Miller and Tucker, "The New Urban Politics," 97.

9. Kaileigh Peyton, "Bobbie Sterne Was Born to Run," *Cincinnati Magazine*, May 2, 2018, https://www.cincinnatimagazine.com/features/bobbie-sterne-was-born-to-run/.

10. Gerald Springer, "Forty-Four Neighborhoods—One Great City," 1973, Springer Papers, US-87-6, Folder 12, BL; Roy Bode, "City Manager: Robert Turner, Impatient Evolutionary at City Hall," *Cincinnati Magazine*, November 1972; Polk Laffoon IV, "How Provincial Are We Really?" *Cincinnati Magazine*, April 1976; Associated Press, "A City Manager Builds a New Life as President of Philadelphia's Zoo," *New York Times*, September 23, 1979; J. Patrick O'Connor, "Interview: William Donaldson," *Cincinnati Magazine*, October 1976; Lee Mitang, "City Manager Stresses Right Decisions," *Spokesman Review* (Spokane, WA), November 27, 1977.

11. Springer, "Announcement of Candidacy—Jerry Springer"; Springer, "Saylor Park," "Jerry Springer's First Term in City Council"; "City Hall in Mt. Washington?" BL.

12. Author's interview with Ben Kaufman, October 16, 2018.

13. City of Cincinnati, "An Ordinance No. 207—1972"; The Citizens Task Force on Environmental Quality, "Report," 1973; "Report of the Citizens

Task Force on Environmental Projection," June 1971, PLCHC; Richard Gibeau, "Environment: A Challenge to City," *Cincinnati Post*, April 5, 1973; Zane L. Miller and Bruce Tucker, *Changing Plans for America's Inner Cities: Cincinnati's Over-the-Rhine and Twentieth-Century Urbanism* (Columbus: The Ohio State University Press, 1998), 98–99.

14. Richard Gibeau, "Sparks Fly at Nuclear Meeting," *Cincinnati Post*, December 15, 1972; Kay Brookshire and Davilynn Furlow, "Energy," *Cincinnati Post*, October 31, 1973.

15. "Leadership Hails Environment Hearings," *Cincinnati Enquirer*, December 17, 1972; Citizens Task Force on Environmental Quality, "Minority Report," 1973, PLCHC.

16. Jose Beduya, "Remembering Physicist Sternglass, Who Helped the World See Man on the Moon," *Cornell Chronicle*, July 17, 2019, https://news.cornell.edu/stories/2019/07/remembering-physicist-sternglass-who-helped-world-see-man-moon; Joseph J. Mangano, "Remembering Ernest Sternglass," *Nation*, March 10, 2015, https://www.thenation.com/article/archive/remembering-ernest-sternglass/.

17. E. J. Sternglass, "Significance of Radiation Monitoring Results for the Shippingport Nuclear Reactor," January 21, 1973, in the Citizens Task Force on Environmental Quality, "Appendix," 1973, PLCHC; Jo-Ann Albers, "River-Cancer Link Probe Sought," *Cincinnati Enquirer*, April 20, 1973; Eugene L. Saenger, "A Survey of Cancer Death Rates in the Ohio Valley," Theodore M. Berry Papers, Box 141, Folder Radioactivity Correspondence and Reports, 1973, BL; Richard Gibeau, "Task Force to Vote on Radiation Proposal," *Cincinnati Post*, April 25, 1973.

18. Ernest J. Sternglass, "Dear Mr. Berry," April 18, 1973, Theodore M. Berry Papers, Box 141, Folder Radioactivity Correspondence and Reports, 1973, BL.

19. Theodore Berry, "The Honorable John J. Gilligan, Dear Sir," April 20, 1973; Mrs. George Matteucci, "Mayor Theodore M. Berry," April 23, 1973, Theodore M. Berry Papers, Box 141, Folder Radioactivity Correspondence and Reports, 1973, BL.

20. NRC, "Supplement to the Summary Report on the Assessment of Environmental Radioactivity in the Vicinity of the Shippingport Power Station," May 31, 1973; Joel Weisman, "Dear Mayor Berry," April 20, 1973; Ira L. Whitman and John W. Cashman, "Dear Mayor Berry," May 29, 1973; Edward C. Pandorf, "Editor, The Cincinnati Enquirer," April 21, 1973; E. Robert Turner, "Gentleman, Re: Radio Active Pollution-Ohio River," July 5, 1973, Theodore M. Berry Papers, Box 141, Folder Radioactivity Correspondence and Reports, 1973, BL.

21. Jo-Ann Albers, "River-Cancer Link Probe Sought," *Cincinnati Enquirer*, April 20, 1973; Walker, *Three Mile Island*, 38–39.

22. Ira L. Whitman and John W. Cashman, "Dear Mayor Berry," May 29, 1973; E. Robert Turner, "Dear Mayor Berry," May 18, 1973; Eugene L.

Saenger, "Dear Sir," May 31, 1973, Theodore M. Berry Papers, Box 141, Folder Radioactivity Correspondence and Reports, 1973, BL.

23. The Citizens Task Force on Environmental Quality, "Report," 1973, PLCHC.

24. The Citizens Task Force on Environmental Quality, "Power Plant Resolution" and "Minority Report," 1973, PLCHC.

25. The Citizens Task Force on Environmental Quality, "Minority Report," 1973, PLCHC.

26. "Turner Opposes City's Entering AEC Hearings," *Cincinnati Enquirer*, September 5, 1973.

27. "W. C. Beckjord, Former Head of CG&E, Is Dead," *Cincinnati Enquirer*, January 1, 1966.

28. "Springer Asks Ban on Nukes," *Cincinnati Enquirer*, February 19, 1976; David Bauer, "Contradictions Bother Councilmen," *Cincinnati Enquirer*, February 19, 1976; Margaret Josten, "Lethal Material Transportation Controls Urged," *Cincinnati Enquirer*, March 23, 1976.

29. "Springer Nuclear Views Protested," *Cincinnati Enquirer*, February 20, 1976.

30. Dennis Doherty, "Nuclear Ban Action Delayed," *Cincinnati Enquirer*, February 24, 1976.

31. Walker, *Three Mile Island*, 7, 10, 132.

32. Walker, *The Road to Yucca Mountain*, 157–160.

33. See Aron, *Licensed to Kill?*; Bedford, *Seabrook Station*; and Wellock, *Critical Masses*.

34. Hirsh, *Power Loss*, 28, 43–45, 170.

35. As the power provider for portions of northern Kentucky and southeast Indiana, CG&E had subsidiaries there, like Union Light, Heat and Power Company in northern Kentucky, but these utilities got rate increases through their respective state public utility commissions.

36. CG&E, "Annual Report," 1973, DE.

37. Hirsh, *Power Loss*, 26–27, 56–58.

38. CG&E, "Statement on Future Gas Supply," May 16, 1972, Berry Papers, Box 141, Folder Cincinnati Gas & Electric Reports and Other Materials, 1972–1975 (Part 2 of 3), BL; "Higher Rates for Utilities," *Cincinnati Post*, June 21, 1972.

39. Mr. and Mrs. Thomas Busemeyer, "Dear Sir," June 15, 1972; "Statement by Councilmen Theodore M. Berry and Gerald N. Springer," June 14, 1972; William A. McClain, "Gentleman," May 8, 1972, Berry Papers, Box 141, Folder Cincinnati Gas & Electric Reports and Other Materials, 1972–1975 (Part 1 of 3); E. Robert Turner, "Gentleman," November 14, 1973, Berry Papers, Box 141, Folder Cincinnati Gas & Electric Reports and Other Materials, 1972–1975 (Part 3 of 3), BL.

40. Frank E. Smith, "Dear Charles," May 19, 1972; Greater Cincinnati Chamber of Commerce, "Projected Economic Growth and the Demand for

Electric Energy in the Cincinnati, Columbus and Dayton Metropolitan Areas 1970 to 1980," February 1972, Berry Papers, Box 141, Folder Cincinnati Gas & Electric Reports and Other Materials, 1972–1975 (Part 2 of 3), BL.

41. "By-Leave Resolution to Be Introduced at City Council, Wednesday, May 2," April 27, 1972; Theodore M. Berry, "Motion," May 3, 1972; CG&E, "Statement on Future Gas Supply" and "Statement on Future Electric Supply," May 16, 1972, Berry Papers, Box 141, Folder Cincinnati Gas & Electric Reports and Other Materials, 1972–1975 (Part 2 of 3), BL.

42. Hirsh, *Power Loss*, 60–61.

43. Kay Brookshire and Davilynn Furlow, "Our Lifestyle Is Running Short of Fuel," *Cincinnati Post*, October 29, 1973; "Energy," *Cincinnati Post*, October 31, 1973; "How Will We Power the Future?" *Cincinnati Post*, November 1, 1973; "City Hall Tightens Its Energy Belt," *Cincinnati Post*, November 8, 1973; Hirsh, *Power Loss*, 61.

44. Borstelmann, *The 1970s*, 53–55; Federal Reserve Bank of St. Louis—Economic Research, "The Great Inflation, 1986–1982," November 22, 2013, https://www.federalreservehistory.org/essays/great-inflation.

45. Walker, *Three Mile Island*, 8.

46. CG&E, "Annual Report," 1975, DE.

47. CG&E, "Annual Report," 1975, DE; Citizen Correspondence against Cincinnati Gas & Electric Rate Increases, 1975, Berry Papers, Box 141, Folder Cincinnati Gas & Electric Reports and Other Materials, 1972–1975 (Part 1 of 3), BL.

48. Meg Jacob, *Panic at the Pump: The Energy Crisis and the Transformation of American Politics in the 1970s* (New York: Hill & Wang, 2017), 41–85; Hirsh, *Power Loss*, 66.

49. Jean C. Renick, "Dear Sir," September 24, 1975, Berry Papers, Box 141, Folder Cincinnati Gas & Electric Reports and Other Materials, 1972–1975 (Part 3 of 3); Charlotte J. Krundieck, "Dear Mayor Berry," October 1, 1975, Berry Papers, Box 141, Folder Citizen Correspondence against Cincinnati Gas & Electric Rate Increases, 1975, BL.

50. CG&E, "Annual Report," 1974, DE; Thomas A. Luebbers, "Special Counsel for CG&E Rate Case," March 25, 1975; Thomas A. Luebbers, "Approved for Submission to Council," July 2, 1975, Berry Papers, Box 141, Folder Cincinnati Gas & Electric Reports and Other Materials, 1972–1975 (Part 2 of 3), BL; "City Seeking Lawyer to Battle CG&E," *Cincinnati Post*, March 26, 1975.

L. Shuttlesworth, Rev. Otis Moss Jr., and Rev. William Ricke,
City," March 6, 1975; Shuttleworth, Otis, and Ricke,
Iuman Events of Our Day . . .," March 3, 1975; City
.—1975," 1975; Thomas A. Luebbers, "Approved for
July 2, 1975, Berry Papers, Box 141, Folder Cincinnati
nd Other Materials, 1972–1975 (Part 3 of 3), BL.
Members of Cincinnati City Council," May 19, 1972;
Columbia Gas of Ohio, Inc., 479 F.2d 153 (6th Cir.

1973), Berry Papers, Box 141, Folder Cincinnati Gas & Electric Reports and Other Materials, 1972–1975 (Part 1 of 3), BL; see Jacob, *Panic at the Pump*.

53. CG&E, "Annual Report," 1975–1979, DE.

4. "Not Another Harrisburg"

1. CG&E, "Annual Report," 1978, DE.
2. Author's interview with David Fankhauser, August 5, 2017.
3. Walker, *Three Mile Island*, 71–79, 190–208.
4. NRC, *Hearing in the Matter of Cincinnati Gas & Electric Company*, November 14, 1979, NRC.
5. Wellock, *Safe Enough*, 75; Walker, *Three Mile Island*, 71–79.
6. *Meltdown: Three Mile Island,* episodes 1–2, "The Accident" and "Women and Children First," directed by Kief Davidson, aired May 10, 2022, on Netflix, https://www.netflix.com/watch/81241561?trackId=14277283&tctx=-97%2C-97%2C%2C%2C%2C%2C%2C%2C81198239.
7. Ibid.
8. NRC, "Population Dose and Health Impact of the Accident at Three Mile Island Nuclear Station," May 1979, DF; *Meltdown: Three Mile Island,* episode 4, "Fallout," directed by Kief Davidson, aired May 10, 2022, on Netflix, https://www.netflix.com/watch/81241561?trackId=14277283&tctx=-97%2C-97%2C%2C%2C%2C%2C%2C%2C81198239.
9. Wellock, *Safe Enough*, 74–76.
10. Steve Wilson, "Ohio Nuclear Safety Gets 'Another Look,'" *Cincinnati Enquirer*, April 3, 1979; "Safety Officials Check A-Plant," *Cincinnati Enquirer*, April 8, 1979; Steve Willis, "Inspection of Zimmer Plant 'Reassures' Ohio Task Force," *Cincinnati Enquirer*, May 2, 1979.
11. NRC, "Safety Evaluation Report," January 1979, DF; "Radiation Monitoring in the Works," *Cincinnati Post*, June 19, 1979; Randy McNutt, "Gradison Will Ask for Zimmer Delay," *Cincinnati Enquirer*, May 30, 1979.
12. NRC, *Hearing in the Matter of Cincinnati Gas & Electric Company*, May 22, 1979, NRC; "Dear Councilman Brush," April 16, 1979; "Dear Councilman Brush," April 18, 1979, "Dear Tom Brush," April 18, 1979, Thomas Brush Papers, MSS 954, Box 34, Folder Zimmer Power Station, April 1979–December 1981, Cincinnati Museum Center History Library and Archives, Cincinnati, OH (hereafter CMC).
13. Dave Krieger, "City to Seek Delay of Hearing on Zimmer License," *Cincinnati Enquirer*, May 3, 1979; "Radiation Monitoring in the Works," *Cincinnati Post*, June 19, 1979; Sharon Moloney, "Council to Ask for Zimmer Delay," *Cincinnati Post*, May 3, 1979. For letters to council, see Thomas Brus[illegible] Papers, MSS 954, Box 34, Folder Zimmer Power Station, April 1979–Dece[illegible] 1981.

14. "The Kemeny Commission's Duty," *New York Times*, April 15, 1979; Walker, *Three Mile Island*, 51–70, 209–215; NRC, "TMI-2 Lessons Learned Task Force Final Report," October 1979; NRC, "TMI-2 Lessons Learned Task Force Status Report and Short-Term Recommendations," July 1979, DF.

15. Walker, *Three Mile Island*, 220–222.

16. CG&E, "Annual Report," 1979, DE; CG&E, "How Safe Is Zimmer?" *Cincinnati Post*, June 18, 1979; Scott Aiken, "CG&E Shareholders Back A-Plant," *Cincinnati Enquirer*, April 26, 1979; Randy McNutt, "Speakers Disagree over Zimmer's Effect on Water," *Cincinnati Enquirer*, April 12, 1979.

17. Robin Herman, "Nearly 200,000 Rally to Protest Nuclear Energy," *New York Times*, September 24, 1979.

18. Douglas Starr, "Anti-Nuclear Vigil Staged Here," *Cincinnati Post*, March 31, 1979; Dave Krieger, "City to Seek Delay of Hearing on Zimmer License," *Cincinnati Enquirer*, May 3, 1979; "3-Day Zimmer Plant Conferences Open Monday," *Cincinnati Enquirer*, May 20, 1979.

19. Douglas Starr, "27 Arrested for Staging Sit-In at Zimmer," *Cincinnati Post*, June 4, 1979.

20. Ibid.; Chew, "Power from the Valley," 127–136.

21. Randy McNutt, "Judge to Set New Trial Date," *Cincinnati Enquirer*, October 27, 1979.

22. Tom Mueller, *Crisis of Conscience: Whistleblowing in an Age of Fraud* (New York: Riverhead, 2019), 272–274.

23. Author's interview with Tom Carpenter, March 3, 2018.

24. Jeremy Suri, *Power and Protest: Global Revolution and the Rise of Détente* (Cambridge, MA: Harvard University Press, 2003), 213–259; Lawrence S. Wittner, *Toward Nuclear Abolition: A History of the World Nuclear Disarmament Movement, 1971–Present* (Stanford, CA: Stanford University Press, 2003), 1–20, 97–106.

25. "5,000 in Colorado Protest a Nuclear Weapons Plant," *New York Times*, April 30, 1978.

26. Author's interview with Tom Carpenter, March 3, 2018.

27. John Eliot, "Polly Speaks," *Cincinnati Post*, April 21, 1973; Mary Gottschalk, "Memorial Service Set for Amos Brokaw," *Mercury News* (San Jose, CA), February 24, 2011; Paul Furgia, "Protester Takes Jail over Fine," *Cincinnati Enquirer*, July 19, 1983; Ben L. Kaufman, "Pacifist Kills Self to Escape Alzheimer's," *Cincinnati Enquirer*, November 12, 1997.

28. Author's interview with Tom Carpenter, March 3, 2018.

29. For more on Fernald, see Carol Rainey, *One Hundred Miles from Home: Nuclear Contamination in the Communities of the Ohio River Valley* (Cincinnati: Cyndell Press, 2008); Steven R. Langois, "Feeding a Nuclear Giant: Fernald and the Uranium Production System, 1943–1989" (MA thesis, University of Saskatchewan, Saskatoon, 2019); Casey A. Huegel, "Fernald and the Transformation of Environmental Activism: The Grassroots Movement to

Make America Safe from Nuclear Weapons Production" (PhD diss., University of Cincinnati, 2022); and Casey Huegel, *Cleaning Up the Bomb Factory: Grassroots Activism and Nuclear Waste in the Midwest* (Seattle: University of Washington Press, 2024).

30. Author's interview with Tom Carpenter, March 3, 2018; CARE, "Let OPEC Keep Their Oil," flyer, DF; CARE, "The Most Dangerous Idea to Ever Hit the Midwest," *Cincinnati Enquirer*, April 15, 1981.

31. Author's interview with Tom Carpenter, March 3, 2018.

32. Ibid.

33. Ibid.

34. John Erardi, "Anti-Nuclear Protesters Knock Area Power Plants," *Cincinnati Enquirer*, October 29, 1978.

35. Mothers for Peace, "Our History Timeline," 2023, https://mothersforpeace.org/about-us/our-history/; see also Barbara Epstein, *Political Protest and Cultural Revolution: Nonviolent Direct Action in the 1970s and 1980s* (Berkeley: University of California Press, 1993).

36. Adam Bernstein, "Charles Bechhoefer, Administrative Judge, Dies at 80," *Washington Post*, August 26, 2013.

37. NRC, *Hearing in the Matter of Cincinnati Gas & Electric Company*, November 14, 1979; NRC, *Hearing in the Matter of Cincinnati Gas & Electric Company*, May 23, 1979, NRC.

38. NRC, *Hearing in the Matter of Cincinnati Gas & Electric Company*, November 14, 1979, NRC.

39. Ibid.; NRC, *Hearing in the Matter of Cincinnati Gas & Electric Company*, May 22, 1979, NRC.

40. NRC, *Hearing in the Matter of Cincinnati Gas & Electric Company*, November 14, 1979, NRC.

41. Ibid.

42. Ibid.

43. Ibid.; NRC, *Hearing in the Matter of Cincinnati Gas & Electric Company*, May 22, 1979, NRC.

44. Ibid.

45. Quoted in Linda Nash, *Inescapable Ecologies: A History of Environment, Disease, and Knowledge* (Berkeley: University of California Press, 2007), 177. See also Elizabeth Blum, *Love Canal Revisited: Race, Class, and Gender in Environmental Activism* (Lawrence: University Press of Kansas, 2011); Phil Brown and Edwin Mikkelson, *No Safe Place: Toxic Waste, Leukemia, and Community Action* (Berkeley: University of California Press, 1997).

46. See Natasha Zaretsky, *Radiation Nation: Three Mile Island and the Political Transformation of the 1970s* (New York: Columbia University Press, 2018) for this reaction. See also NRC, "Population Dose and Health Impact of the Accident at Three Mile Island Nuclear Station," May 1979, DF; *Meltdown: Three Mile Island*, episode 4, "Fallout."

47. See Willis, *Conservation Fallout*, chap. 3, for more on Mothers for Peace.

48. NRC, *Hearing in the Matter of Cincinnati Gas & Electric Company*, November 14, 1979, NRC.

49. Author's interview with Alice Gerdeman, July 18, 2014; Patricia Gallagher Newbury, "Dead End," *City Beat*, November 1, 2011; Saad Ghosn, "A Natural Path for Peace and Justice: Sister Alice Gerdeman's Faith Shows Her the Way," *sos art cincinnati*, accessed April 12, 2023, https://sosartcincinnati.com/2018/03/23/a-natural-path-for-peace-justice-sister-alice-gerdemans-faith-shows-her-the-way/.

50. NRC, *Hearing in the Matter of Cincinnati Gas & Electric Company*, November 14, 1979, NRC.

51. Ben L. Kaufman, "Colleagues Give Honor to Attorney," *Cincinnati Enquirer*, May 26, 1993; Sharon Morgan, "Andrew Dennison Argued Controversial Cases in Court," *Cincinnati Enquirer*, June 13, 1993.

52. Author's interview with Alice Gerdeman, July 18, 2014; "Kentucky Asks Say in Zimmer Hearings," *Cincinnati Post*, March 5, 1980; Vicky Anderson Mayer and Nancy Flaherty, "The Zimmer Debate: 'What We Learned Has Frightened Us,'" *Cincinnati Post*, August 9, 1979.

53. Ben L. Kaufman, "Clermont Residents Leery of Nuclear Disaster Plan," *Cincinnati Enquirer*, May 4, 1979; Douglas Starr, "Movement Grows for Zimmer Delay," *Cincinnati Post*, May 1, 1979; "3-Day Zimmer Plant Conferences Open Monday," *Cincinnati Enquirer*, May 20, 1979.

54. NRC, *Hearing in the Matter of Cincinnati Gas & Electric Company*, May 22, 1979; NRC, *Hearing in the Matter of Cincinnati Gas & Electric Company*, May 23, 1979, NRC.

55. NRC, *Hearing in the Matter of Cincinnati Gas & Electric Company*, May 22, 1979, NRC.

56. Jefferson Cowie, *Stayin' Alive: The 1970s and the Last Days of the Working Class* (New York: The New Press, 2010), 12.

57. Author's interview with David Fankhauser, August 5, 2017.

58. See Zaretsky, *Radiation Nation*.

59. Douglas Starr, "Zimmer Hearings Focus on Students," *Cincinnati Post*, June 20, 1979.

60. Ibid.

61. NRC, "Draft Environmental Statement Related to the Operation of Wm. H. Zimmer Nuclear Power Plant," October 1976, DF.

62. Ibid.

63. NRC, "Safety Evaluation Report," January 1979, DF.

64. Douglas Starr, "Zimmer Hearings Focus on Students," *Cincinnati Post*, June 20, 1979.

65. Author's interview with David Fankhauser, August 5, 2017; NRC, "Safety Evaluation Report," January 1979, DF; Charles Bechhoefer, "Order Admitting New Contentions and Establishing Discovery Schedule with Regard

Thereto," April 10, 1979, DF; Douglas Starr, "Studies Will Delay Decision on Zimmer," *Cincinnati Post,* May 24, 1979; Ben L. Kaufman, "Delay Boosts Zimmer Cost $186 Million," *Cincinnati Enquirer*, August 1, 1979.

66. Bob McKay, "The Tug of War over Nuclear Energy," *Cincinnati Magazine*, August 1977.

67. NRC, *Hearing in the Matter of Cincinnati Gas & Electric Company*, November 14, 1979, NRC.

68. Ibid.; NRC, *Hearing in the Matter of Cincinnati Gas & Electric Company*, May 22, 1979, NRC.

69. Ibid.

70. NRC, *Hearing in the Matter of Cincinnati Gas & Electric Company*, May 22, 1979, NRC.

71. Borstelmann, *The 1970s,* 46. For more on declining public trust in the 1970s, see also Beth Bailey and David Farber, eds., *America in the Seventies* (Lawrence: University Press of Kansas, 2004); Borstelmann, *The 1970s*; Philip Jenkins, *Decade of Nightmares: The End of the Sixties and the Making of the Eighties* (New York: Oxford University Press, 2006); Kevin M. Kruse and Julian E. Zelizer, *Fault Lines: A History of the United States Since 1974* (New York: W. W. Norton, 2019), 7–43; Bruce J. Schulman, *The Seventies: The Great Shift in American Culture, Society, and Politics* (New York: The Free Press, 2001).

72. Paul Sabin, *Public Citizens: The Attack on Big Government and the Remaking of American Liberalism* (New York: W. W. Norton, 2021), 3–32.

73. Borstelmann, *The 1970s,* 19–72.

74. Zaretsky, *Radiation Nation*, 127.

5. "The Dirtiest Plant I've Ever Seen"

1. Author's interview with Tom Devine, December 13, 2019; Mary Carmen Cupito, "Investigator Recounts Plant Probe," *Cincinnati Post*, November 25, 1981.

2. Thomas Devine, "Dear Mr. Keppler," May 11, 1981, Thomas Brush Papers, Box 34, MSS 954, Folder Zimmer Power Station, April 1979–December 1981, CMC.

3. Ibid.

4. Mary Carmen Cupito, "Investigator Recounts Plant Probe," *Cincinnati Post*, November 25, 1981.

5. Author's interview with Tom Carpenter, March 3, 2018.

6. Ibid.

7. Author's interview with Tom Devine, December 13, 2019.

8. "Nuclear Plants' Problems Said to Rise: 'Minor to Very Serious,'" *New York Times*, July 28, 1981.

9. Author's interview with Tom Carpenter, March 3, 2018.

10. See Sabin, *Public Citizens*, particularly part 2.

11. Author's interview with Tom Devine, December 13, 2019.

12. Author's interview with Tom Carpenter, March 3, 2018.

13. Author's interview with Phil Amadon, June 25, 2014; Government Accountability Project, "Tom Devine," accessed June 8, 2022, https://whistleblower.org/our-team/tom-devine/.

14. Author's interview with Tom Devine, December 13, 2019.

15. Quoted in Barbara Redding, "CG&E 'OK' on Zimmer Inspections," *Cincinnati Enquirer*, February 27, 1976; Lawrence Sussman, "Zimmer Engineer Who Quit Rejects Findings," *Cincinnati Post*, February 27, 1976.

16. "3 Zimmer Complaints Checked," *Cincinnati Post*, February 9, 1979; Jim Greenfield, "After Slow Start, Latest Testimony Bolsters Zimmer's Foes," *Cincinnati Enquirer*, July 1, 1979.

17. Douglas Starr, "Affidavit: Zimmer Welds Bad," *Cincinnati Post*, June 29, 1979.

18. Douglas Starr, "CG&E Challenges Worker's View That Zimmer Reactor Seals Faulty," *Cincinnati Post*, August 9, 1979; "Zimmer's Faulty Doors Fixed," *Cincinnati Post*, June 11, 1979.

19. "Zimmer Inspector Fired; Protests Await Fuel," *Cincinnati Post*, August 15, 1979; Douglas Starr, "Affidavit: Zimmer Welds Bad," *Cincinnati Post*, June 29, 1979; "A-Plant Allegations Represent 'Fantasy,'" *Cincinnati Enquirer*, November 16, 1979; "Power Clash," *Cincinnati Post*, April 12, 1979.

20. Douglas Starr, "Zimmer Opening May Be Delayed," *Cincinnati Post*, February 15, 1979.

21. NRC, *Hearing in the Matter of Cincinnati Gas & Electric Company*, August 10, 1979, NRC.

22. Douglas Starr, "Allegation Highlights Zimmer Hearing," *Cincinnati Post*, June 22, 1979.

23. NRC, *Hearing in the Matter of Cincinnati Gas & Electric Company*, November 15, 1979, NRC.

24. Ibid.

25. Ibid.

26. NRC, *Hearing in the Matter of Cincinnati Gas & Electric Company*, August 10, 1979; NRC, *Hearing in the Matter of Cincinnati Gas & Electric Company*, November 14, 1979, NRC.

27. Quoted in Zaretsky, *Radiation Nation*, 85; Walker, *Permissible Dose*, 92–93.

28. NRC, *Hearing in the Matter of Cincinnati Gas & Electric Company*, May 23, 1979, NRC; author's interview with Stan Hedeen, October 23, 2019.

29. Author's interview with Tom Devine, December 13, 2019.

30. Thomas Devine, "Dear Mr. Keppler," May 11, 1981, Thomas Brush Papers, Box 34, MSS 954, Folder Zimmer Power Station, April 1979–December 1981, CMC.

31. Ibid.

32. Ibid.

33. Ibid.

34. Ibid.

35. Ibid.

36. Ibid.

37. Ibid.

38. Ibid.

39. Ibid.

40. Author's interview with Tom Devine, December 13, 2019.

41. Devine, "Dear Mr. Keppler."

42. Ben L. Kaufman, "NRC Says It Bungled Plant Probe," *Cincinnati Enquirer*, November 18, 1981; Joanne Omang, "NRC Faults Its Own Safety Probe at Ohio A-Plant," *New York Times*, November 21, 1981; Thomas Devine, "Testimony of Tom Devine on Behalf of the Government Accountability Project of the Institute for Policy Studies," June 10, 1982, Marian and Donald Spencer Papers, US 91-2, Box 2, Folder 67, Nuclear Freeze, BL.

43. Author's interview with Tom Devine, December 13, 2019.

44. Mary Carmen Cupito, Robert White, and Ron Liebau, "CG&E Fined $200,000 for Zimmer," *Cincinnati Post*, November 25, 1981.

45. Author's correspondence with Ben Kaufman, February 26, 2023; Ben L. Kaufman, "Zimmer Incurs $200,000 Fine over Foul-Ups," *Cincinnati Enquirer*, November 26, 1981; Mary Carmen Cupito, Robert White, and Ron Liebau, "CG&E fined $200,000 for Zimmer," *Cincinnati Post*, November 25, 1981.

46. Author's interview with Tom Devine, December 13, 2019.

47. "2,300 Incidents Reported at Nuclear Plants in 1979," *New York Times*, July 14, 1980; "Nuclear Plants' Problems Said to Rise: 'Minor to Very Serious,'" *New York Times*, July 28, 1981.

48. Matthew L. Wald, "U.S. Study Reassesses Risk of Nuclear Plant Accidents," *New York Times*, July 6, 1982.

49. Author's interview with Stan Hedeen, October 23, 2019. For more on all these issues, see Wellock, *Safe Enough*, 78.

50. Chew, "Power from the Valley," 140–142.

51. John O'Dell, "Charges by Fired Engineer: Allegations of Substandard Welding Raised at Onofre," *New York Times*, October 13, 1982.

52. See Willis, *Conservation Fallout*, chap. 3.

53. Bedford, *Seabrook Station*, 116–120, 145; *Meltdown: Three Mile Island*, episode 4, "Fallout," directed by Kief Davidson, aired May 10, 2022, on Netflix, https://www.netflix.com/watch/81241561?trackId=14277283&tctx=-97%2C-97%2C%2C%2C%2C%2C%2C%2C81198239.

54. Author's interview with Tom Carpenter, March 3, 2018; Mueller, *Crisis of Conscience*, chap. 5. See also Huegel, "Fernald and the Transformation of Environmental Activism," especially chapter 3 on Fernald workers.

55. "2,300 Incidents Reported at Nuclear Plants in 1979," *New York Times*, July 14, 1980. See Aron, *Licensed to Kill* for more on public distrust toward the NRC.

56. Ben L. Kaufman, "Zimmer Scrutiny Ordered by NRC," *Cincinnati Enquirer*, November 21, 1981.

57. Ron Liebau, "Zimmer Polishes Image," *Cincinnati Post*, December 9, 1980; author's correspondence with Ben Kaufman, February 16, 2023.

58. Wellock, *Safe Enough*, 74.

59. Dan Andriacco, "Zimmer Costs Now $1 Billion," *Cincinnati Post*, April 24, 1980.

6. "It's Time to Stop Bailing the Utilities Out"

1. CARE, "The Zimmer Nuclear Plant: A Risk Cincinnati Can't Afford," *Cincinnati Enquirer*, March 28, 1980.

2. CG&E, "Annual Report," 1981, DE.

3. CG&E, "Annual Report," 1978, 1979, 1980–81, DE.

4. CG&E, "Annual Report," 1979, 1978, DE.

5. Felicia Lee, "First Neighborhood Conference Produces Encouraging Results," *Cincinnati Enquirer*, September 28, 1980; Ron Rollins, "Utility Reform Group Wins a Few Rounds," *Cincinnati Enquirer*, May 31, 1981.

6. Sherri Blank, "Protesters Register Their Displeasure at Hearing on CG&E Rate Hike Request," *Cincinnati Enquirer*, December 12, 1980; Felicia Lee, "First Neighborhood Conference Produces Encouraging Results," *Cincinnati Enquirer*, September 28, 1980; Ben L. Kaufman, "Catholic Money Fueling Challenge to CG&E Rates," *Cincinnati Enquirer*, October 22, 1980.

7. LaVerne Willsey in 1940 and 1950 Federal Censuses, ancestry.com; Jim Sluzewski, "Persistence, Unity Pay Off in Besieging City Hall," *Cincinnati Enquirer*, February 18, 1980; Mark Kestigian, "Northside Group Ready to Square Off with CG&E," *Cincinnati Post*, August 25, 1978; Larry Bivins, "Belated Celebration: Fuel Adjustment Revision Toasted," *Cincinnati Post*, August 19, 1980.

8. "Getting Heard at CG&E," *Cincinnati Magazine*, December 1978.

9. Len Penix, "Feisty Audience Protests Proposed Rise in Utility Bills," *Cincinnati Post*, October 29, 1980; "300 Protest CG&E," *Cincinnati Post*, February 2, 1979; Sherri Blank, "Protesters Register Their Displeasure at Hearing on CG&E Rate Hike Request," *Cincinnati Enquirer*, December 12, 1980.

10. "Getting Heard at CG&E," *Cincinnati Magazine*, December 1978; author's interview with Phil Amadon, June 25, 2014.

11. "City Readies Opposition to CG&E Rate Request," *Cincinnati Post*, November 1, 1979; Larry Bivins, "Belated Celebration: Fuel Adjustment Revision Toasted," *Cincinnati Post*, August 19, 1980.

12. Office of the Ohio Consumers' Counsel, "About OCC," Ohio Consumers' Counsel, accessed June 20, 2022, https://www.occ.ohio.gov/about-occ.

13. Larry Bivins, "Belated Celebration: Fuel Adjustment Revision Toasted," *Cincinnati Post*, August 19, 1980.

14. Author's interview with Brewster Rhoads, March 1, 2018.

15. Ibid.

16. Ibid.

17. Ibid.; author's interview with Roxanne Qualls, March 14, 2018.

18. Author's interview with Roxanne Qualls, March 14, 2018.

19. "Zimmer," *Cincinnati Enquirer*, November 28, 1981; author's interview with Brewster Rhoads, March 1, 2018; author's interview with Roxanne Qualls, March 14, 2018.

20. See Lizabeth Cohen, *A Consumer's Republic: The Politics of Mass Consumption in Postwar America* (New York: Alfred A. Knopf, 2003), 345–397.

21. Sabin, *Public Citizens*, 173–175.

22. Hirsh, *Power Loss*, 172; see also Bedford, *Seabrook Station*.

23. Hirsh, *Power Loss*, 171; Chew, "Power from the Valley," 144.

24. Douglas Starr, "CG&E Shares U.S. Nuclear Damage Costs," *Cincinnati Post*, May 4, 1979.

25. NRC, *Hearing in the Matter of the Cincinnati Gas & Electric Company*, March 4, 1981, NRC.

26. Ibid.

27. Ibid.; Ben L. Kaufman, "Eventual Razing of Zimmer Plant May Cost Billions," *Cincinnati Enquirer*, March 9, 1981.

28. NRC, *Hearing in the Matter of the Cincinnati Gas & Electric Company*, March 4, 1981, NRC.

29. NRC, "Low-Level Waste Compacts," March 12, 2020, https://www.nrc.gov/waste/llw-disposal/licensing/compacts.html. Ultimately, after the Midwest compact was official in 1985, no official site was established within those states. As of 2020, the US has four such sites—in Nevada, South Carolina, Texas, and Washington—none of which serves the Midwest compact.

30. Ben L. Kaufman, "Eventual Razing of Zimmer Plant May Cost Billions," *Cincinnati Enquirer*, March 9, 1981.

31. Ibid.

32. NRC, *Hearing in the Matter of the Cincinnati Gas & Electric Company*, March 3, 1981, NRC; Federal Reserve Bank of St. Louis—Economic Research, "The Great Inflation, 1986–1982," November 22, 2013, https://www.federalreservehistory.org/essays/great-inflation.

33. Eliot Marshall, "Utilities Lose Power on Wall Street," *Science*, January 30, 1981.

34. Lifset, *Power on the Hudson*, 10; Bedford, *Seabrook Station*, 112.

35. Hirsh, *Power Loss*, 171–172, 179, 161; Wellock, *Safe Enough*, 77; CG&E, "Annual Report," 1978, DE.

36. NRC, *Hearing in the Matter of the Cincinnati Gas & Electric Company*, March 4, 1981, NRC.

37. CG&E, "Annual Report," 1980, DE.

38. Hirsh, *Power Loss*, 172.

39. Ibid., 75–131.

40. Ibid., 135–167.

41. Ibid., 167, 182; quoted in Richard Munson, *The Power Makers: The Inside Story of America's Biggest Business—and Its Struggle to Control Tomorrow's Electricity* (Emmaus, PA: Rodale Press, 1985), 182.

42. NRC, *Hearing in the Matter of the Cincinnati Gas & Electric Company*, March 4, 1981, NRC.

43. Quoted in ibid.

44. Bedford, *Seabrook Station*, 111; NRC, *Financial Qualifications Statement of Policy*, June 14, 1984, https://www.nrc.gov/docs/ML2009/ML20092D900.pdf.

45. Author's interview with Brewster Rhoads, March 1, 2018.

46. Jackie Jadrnak, "Worker Claims CG&E Wasteful at Zimmer," *Cincinnati Enquirer*, December 10, 1981.

47. "Consumer Lawyer Asks for Zimmer Audit," *Cincinnati Enquirer*, December 12, 1981.

48. CG&E, "Annual Report," 1981, DE.

7. "It Was Like a Dam Breaking"

1. Ben L. Kaufman, "Utilities Will Pay $200,000 Fine for Zimmer Foul-Ups," *Cincinnati Enquirer*, February 25, 1982.

2. Quoted in Bedford, *Seabrook Station*, 142.

3. Ben L. Kaufman, "Simulated Radiation Spreads in Zimmer 'Crisis,'" *Cincinnati Enquirer*, November 19, 1981; Mary Carmen Cupito, Ron Liebau, and Robert White, "Mock Accident Tests Emergency Plans at Zimmer," *Cincinnati Post*, November 18, 1981; Rolf Wiegand, "Many Citizens Unaware Mock Disaster at Zimmer Was Being Conducted," *Cincinnati Enquirer*, November 19, 1981.

4. Ben L. Kaufman, "Simulated Radiation Spreads in Zimmer 'Crisis,'" *Cincinnati Enquirer*, November 19, 1981; "Zimmer Drill Gets Mixed Reviews," *Cincinnati Post*, November 19, 1981.

5. Jeff Gutsell, "KY Town Prevails, Gets Feds to Hold Up Zimmer License," *Cincinnati Enquirer*, June 27, 1982.

6. Ibid.

7. Ron Liebau, "Hearings Put Zimmer Fate on Seesaw," *Cincinnati Post*, July 28, 1982.

8. NRC, *Hearing Before the Atomic Safety and Licensing Appeal Board on the Cincinnati Gas & Electric Company*, January 10, 1983, NRC; NRC, *In the Matter of the Cincinnati Gas & Electric Co, et al.*, June 21, 1982, DF; "Zimmer Plan Has New Routes," *Cincinnati Enquirer*, March 2, 1982; Ron Liebau, "CG&E Evacuation Estimates Conflict," *Cincinnati Post*, October 1, 1980.

9. NRC, *Hearing Before the Atomic Safety and Licensing Appeal Board on the Cincinnati Gas & Electric Company*, January 10, 1983, NRC; NRC, *In the Matter of the Cincinnati Gas & Electric Co, et al.*, June 21, 1982, DF.

10. Author's interview with Alice Gerdeman, July 18, 2014.

11. Ibid.

12. Ibid.

13. NRC, *In the Matter of the Cincinnati Gas & Electric Co, et al.*, June 21, 1982, DF.

14. NRC, *Hearing Before the Atomic Safety and Licensing Appeal Board on the Cincinnati Gas & Electric Company*, January 10, 1983, NRC; NRC, *In the Matter of the Cincinnati Gas & Electric Co, et al.*, June 21, 1982, DF.

15. Author's interview with Alice Gerdeman, July 18, 2014.

16. Bedford, *Seabrook Station*, 125–161; Aron, *Licensed to Kill*, 44–89.

17. Ron Liebau, "Zimmer Clears Major Hurdle," *Cincinnati Post*, April 29, 1982.

18. Aron, *Licensed to Kill*, 56.

19. Ron Liebau, "Hearings Put Zimmer Fate on Seesaw," *Cincinnati Post*, July 28, 1982; Howard Wilkinson, "Emergency Plans Could Delay Zimmer Hearings," *Cincinnati Enquirer*, November 24, 1982; "NRC Report Lauds Zimmer Operators," *Cincinnati Enquirer*, January 27, 1982.

20. David Shapiro and Richard Whitmire, "Washington Group Charges That CG&E Lacks Integrity," *Cincinnati Enquirer*, July 8, 1982.

21. Ibid.

22. Ron Liebau, "Zimmer Problems 'Hidden,'" *Cincinnati Post*, December 10, 1982; author's interview with Tom Devine, December 13, 2019.

23. Liebau, "Zimmer Problems 'Hidden'"; author's interview with Tom Devine, December 13, 2019.

24. Thomas Devine, "Testimony of Tom Devine on Behalf of the Government Accountability Project of the Institute for Policy Studies," June 10, 1982; Thomas Devine, "Introductory Statement for the Briefing of the NRC Commissioners," June 16, 1982, Marian and Donald Spencer Papers, US 91-2, Box 2, Folder 67, Nuclear Freeze, BL.

25. Devine, "Testimony of Tom Devine on Behalf of the Government Accountability Project of the Institute for Policy Studies"; Devine, "Introductory Statement for the Briefing of the NRC Commissioners."

26. Devine, "Testimony of Tom Devine on Behalf of the Government Accountability Project of the Institute for Policy Studies"; Devine, "Introductory Statement for the Briefing of the NRC Commissioners."

27. Devine, "Introductory Statement for the Briefing of the NRC Commissioners"; author's interview with Tom Devine, December 13, 2019.

28. Devine, "Testimony of Tom Devine on Behalf of the Government Accountability Project of the Institute for Policy Studies"; Devine, "Introductory Statement for the Briefing of the NRC Commissioners."

29. Devine, "Introductory Statement for the Briefing of the NRC Commissioners."

30. David Shapiro and Richard Whitmire, "Zimmer Welders Retested," *Cincinnati Enquirer*, July 27, 1982; Jackie Jadrnak, "NRC Official Says Zimmer Startup Unlikely for '83," *Cincinnati Enquirer*, July 15, 1982; Ron Liebau, "Zimmer Inspections Worse, Not Better, Critics Charge," *Cincinnati Post*, October 18, 1982.

31. Thomas Devine, "Supplement to MVPP August 20 Petition to Suspend Construction of the Zimmer Station," October 18, 1982, Thomas Brush Papers, MSS 954, Folder Zimmer Power Station, January–October 1982, CMC.

32. Devine, "Supplement to MVPP August 20 Petition to Suspend Construction of the Zimmer Station"; Thomas Devine, "Summary of Evidence in August 20, MVPP Petition to Suspend Construction of the Zimmer Station," October 18, 1982, Thomas Brush Papers, MSS 954, Folder Zimmer Power Station, January–October 1982, CMC.

33. Camilla Warrick, "NRC Decides against New Hearings on Zimmer," *Cincinnati Enquirer*, August 2, 1982; Ron Liebau, "Zimmer Hearings Blocked," *Cincinnati Post*, July 31, 1982; CG&E, "Annual Report," 1982, DE.

34. Camilla Warrick, "Skeptical CG&E Undecided about Zimmer City Hearings," *Cincinnati Enquirer*, August 11, 1982; author's interview with Brewster Rhoads, March 1, 2018; Cincinnati Environmental Advisory Council, "Special Meeting Minutes," July 14, 1982, Thomas Brush Papers, MSS 954, Folder Zimmer Power Station, January–October 1982, CMC.

35. Mary Carmen Cupito, "'Mystery' Witnesses Testify," *Cincinnati Post*, September 24, 1982; Robert Acomb, "Dear Dave," September 10, 1982, Thomas Brush Papers, MSS 954, Folder Zimmer Power Station, January–October 1982, CMC.

36. "City Council Offers Quality-Control Advice to Nuclear Agency," *Cincinnati Enquirer*, April 15, 1982; Steven Rosen, "Supporters of Arms Freeze to Turn Their Eyes on City," *Cincinnati Enquirer*, August 2, 1982; Tom Carpenter, "Comments on the EAC Zimmer Recommendation," April 14, 1982; D. David Altman, "Report on the Findings and Recommendations of the City of Cincinnati Environmental Advisory Council," October 20, 1982, Thomas Brush Papers, MSS 954, Folder Zimmer Power Station, January–October 1982, CMC.

37. Author's interview with Tom Devine, December 13, 2019.

38. Howard Wilkinson, "Zimmer Subject of Questioning by Grand Jury," *Cincinnati Enquirer*, November 12, 1982; Howard Wilkinson, "Agency Orders Halt to Work at Zimmer," *Cincinnati Enquirer*, November 13, 1982.

39. Marilyn Dillon and Howard Wilkinson, "Shutdown Leaves Zimmer Foes Jubilant," *Cincinnati Enquirer*, November 13, 1982; Ron Liebau, "500 Construction Workers Laid Off," *Cincinnati Enquirer*, October 26, 1982; James Harlow, "Workers Take Pride in Zimmer Quality," *Cincinnati Enquirer*, August 1, 1982.

40. Ron Liebau, "Zimmer Problems 'Hidden,'" *Cincinnati Post*, December 10, 1982.

41. Howard Wilkinson, "From Mule Team, Bechtel Became International Giant," *Cincinnati Enquirer*, November 18, 1982.

42. William Greider, "The Boys from Bechtel," *Rolling Stone*, September 2, 1982; author's interview with Tom Devine, December 13, 2019.

43. Tom Carpenter, "Dear Friend," June 30, 1982, Marian and Donald Spencer Papers, US 91-2, Box 2, Folder 67, Nuclear Freeze, BL.

44. Author's interview with Phil Amadon, June 25, 2014; author's interview with Tom Devine, December 13, 2019.

45. Author's interview with Phil Amadon, June 25, 2014.

46. Ibid.

47. Marilyn Dillon, "CG&E Suffers Setbacks," *Cincinnati Enquirer*, November 16, 1982; "Zimmer Partners Placed on S&P's Creditwatch List," *Cincinnati Enquirer*, December 4, 1982; Ron Liebau, "6-Month Halt Solves Little," *Cincinnati Post*, May 17, 1983.

48. Phil Amadon, "CASE Members," undated, Coalition for Affordable and Safe Energy (hereafter CASE) Papers, personal collection of Phil Amadon, Cincinnati, OH (hereafter PA).

49. Ibid.

50. Ibid.; author's interview with Phil Amadon, June 25, 2014; CASE, "Dear Cincinnati City Council Candidate," September 25, 1983, Marian A. Spencer Papers, MSS 888, Box 9, Folder 44, 1983, 1985, CMC; Rick Anderson, Miami Valley Power Project, "To Mr. Wozniak," August 2, 1983, CASE Papers, PA.

51. Author's interview with Phil Amadon, June 25, 2014.

52. Ibid.; CASE Steering Committee, "Meeting Minutes," May 16, 1983; CASE, "Coalition for Affordable, Safe Energy (C.A.S.E.) Position Statement," undated, CASE Papers, PA. For more on diverse local antinuclear coalitions, see Aron, *Licensed to Kill*, and Bedford, *Seabrook Station*.

53. Chew, "Power from the Valley," 166; CASE, "The C.A.S.E. Steering Committee Unanimously Recommends That the Following Position Be Adopted by C.A.S.E.," undated; Phil Amadon, Rob Dubrow, Jean Donohue, and Rich Reiter, "Proposed Statement of Purpose," February 22, 1984; CASE Steering Committee, "Meeting Minutes," June 15, 1983, CASE Papers, PA.

54. Author's interview with Phil Amadon, June 25, 2014; Bill Gradison, "Letter to Kit Wood," January 5, 1983, CASE Papers, PA; Kit Wood, "Dear

Councilman Brush," October 14, 1982, Thomas Brush Papers, MSS 954, Folder Zimmer Power Station, January–October 1982, CMC.

55. Author's interview with Phil Amadon, June 25, 2014.

56. Archives and Rare Books, University of Cincinnati Libraries, "Marian Spencer: Fighting for Equality in Cincinnati," updated 2022, accessed July 7, 2022, https://libapps.libraries.uc.edu/exhibits/marian-spencer/biography/.

57. Marian Spencer, "Testimony before the Nuclear Regulatory Commission," November 1, 1983, CASE Papers, PA.

58. Willis, *Conservation Fallout*, 100. Willis argues that the antinuclear movement was mostly white because nuclear power was a remote danger compared with the everyday experiences of people of color.

59. Quoted in *Meltdown: Three Mile Island*, episode 2, "Women and Children," directed by Kief Davidson, aired May 10, 2022, on Netflix, https://www.netflix.com/watch/81241561?trackId=14277283&tctx=-97%2C-97%2C%2C%2C%2C%2C%2C%2C81198239; "Nuclear Opposition Rising, Poll Shows," *Cincinnati Enquirer*, May 4, 1982.

60. Author's interview with Tom Devine, December 13, 2019.

Conclusion

1. "CG&E Faces Hearing on Zimmer Review," *Cincinnati Enquirer*, March 28, 1983; Howard Wilkinson, "NRC Okays Torrey Pines as Zimmer Reviewer," *Cincinnati Enquirer*, April 16, 1983.

2. David Shapiro, "Errors by NRC Charged," *Cincinnati Enquirer*, April 20, 1983; Howard Wilkinson, "NRC Ousts Investigations Chief Who Had Guided Zimmer Probe," *Cincinnati Enquirer*, September 30, 1983.

3. Ron Liebau, "Partners Question Zimmer Cost, Delay," *Cincinnati Post*, February 17, 1983.

4. Georgene Kaleina, "Dayton Utility Shareholder Sues CG&E over Zimmer," *Cincinnati Enquirer*, June 3, 1983; Howard Wilkinson, "NRC Awaiting Utility's Proposals on Zimmer," *Cincinnati Enquirer*, August 24, 1983.

5. CG&E, "We Pledge to Complete the Zimmer Nuclear Power Plant and Operate It Efficiently and Safely under the Guidelines of the Nuclear Regulatory Commission," *Cincinnati Enquirer*, December 4, 1982; CARE, "Time To Dismantle CG&E's Big Mistake," *Cincinnati Enquirer*, January 4, 1983; Howard Wilkinson, "Zimmer Foes Launch Fund-Raising Drive," *Cincinnati Enquirer*, December 29, 1982.

6. GAP, "News Release," July 13, 1983, Thomas Brush Papers, MSS 954, Folder Zimmer Power Station, 1983, CMC; Stephanie Jones, "New Charges Critical of Zimmer," *Cincinnati Post*, July 13, 1983.

7. John Eckberg, "Report Blames CG&E for Zimmer Woes," *Cincinnati Enquirer*, August 23, 1983.

8. Torrey Pines Technology, "Independent Review of Zimmer Project Management: Final Report," 6–7, August 1983, PLCHC.

9. John Eckberg, "Report Blames CG&E for Zimmer Woes," *Cincinnati Enquirer*, August 23, 1983; James Hannah, "Zimmer Review Is Seen Taking a Year or More," *Cincinnati Enquirer*, September 29, 1983.

10. Marian Spencer, "Statement by Marian Spencer before the Intergovernmental Affairs Committee," October 10, 1983, Marian A. Spencer Papers, MSS 888, Box 27, Folder 33, William H. Zimmer Nuclear Power Station: Spencer Statements and Press Releases, 1983–84; City Council, "Resolution – 198_," undated; Arnold Bortz, "Gentlemen," December 12, 1983; Ohio Consumers' Counsel, "For Immediate Release," January 20, 1984, Marian A. Spencer Papers, MSS 888, Box 27, Folder 23, William H. Zimmer Nuclear Power Station: Council Resolutions, 1983–84, CMC.

11. Howard Wilkinson, "Managers of Zimmer Hold Talks," *Cincinnati Enquirer*, September 1, 1983; Ron Liebau, "CG&E Hires Consultants for Zimmer," *Cincinnati Post*, June 30, 1983; author's interview with Phil Amadon, June 25, 2014; "Ex-Admiral Is Named to Run Zimmer Plant," *Cincinnati Enquirer*, March 19, 1983; Al Andry, "New Zimmer Chief Says It's 'Big Job,'" *Cincinnati Post*, March 19, 1983.

12. Author's interview with Phil Amadon, June 25, 2014.

13. Author's interview with Tom Devine, December 13, 2019; Associated Press, "Nader Group Calls Zimmer Worst Plant," *Cincinnati Enquirer*, September 20, 1983.

14. Randy Ludlow, "CG&E Says Zimmer to Cost $3.1 Billion," *Cincinnati Post*, October 5, 1983; author's interview with Tom Devine, December 13, 2019.

15. "Mounting Zimmer Cost Worries Guckenberger," *Cincinnati Post*, September 9, 1983; Hirsh, *Power Loss*, 175.

16. Thomas Devine, "Dear Commissioners," December 14, 1983, Marian A. Spencer Papers, MSS 888, Box 27, Folder 29, William H. Zimmer Nuclear Power Station: Government Accountability Project, 1983–84, CMC.

17. Ibid.

18. Ibid.; James G. Keppler, "Gentlemen, to MVPP," January 3, 1984; MVPP, "Dear Mr. Keppler," December 6, 1983, Marian Spencer Papers, MSS 888, Box 27, Folder 34, William H. Zimmer Nuclear Power Station: US Nuclear Regulatory Commission, 1982–85, CMC.

19. CASE, "Purpose of This Hearing," November 1, 1983, Marian Spencer Papers, MSS 888, Box 27, Folder 34, William H. Zimmer Nuclear Power Station: US Nuclear Regulatory Commission, 1982–85, CMC.

20. Howard Wilkinson, "New Zimmer Problems Pop Up before Old Ones Can Be Solved," *Cincinnati Enquirer,* November 2, 1983; author's interview with Phil Amadon, June 25, 2014; Marian Spencer, "Testimony before the Nuclear Regulatory Commission," November 1, 1983, Marian A. Spencer

Papers, MSS 888, Box 27, Folder 33, William H. Zimmer Nuclear Power Station: Spencer Statements and Press Releases, 1983–84, CMC; Jim Joseph, "Prayer, Praise, Ire for Zimmer," *Cincinnati Post*, November 2, 1983.

21. NRC, *Hearing in the Matter of Cincinnati Gas & Electric Company*, December 9, 1983, NRC.

22. Ibid.

23. Ibid.

24. Howard Wilkinson, "Despite Uncertainty, CG&E Files Plan to Finish Zimmer," *Cincinnati Enquirer*, December 28, 1983.

25. Thomas Devine, "Dear Mr. Keppler," October 31, 1983, Thomas Brush Papers, MSS 954, Folder Zimmer Power Station, 1983, CMC.

26. Thomas Devine, "Dear Mr. Keppler," January 11, 1984; Thomas Devine, "Dear Commissioners," November 30, 1983; Thomas Devine, "Dear Commissioners," December 14, 1983, Marian A. Spencer Papers, MSS 888, Box 27, Folder 29, William H. Zimmer Nuclear Power Station: Government Accountability Project, 1983–1984, CMC.

27. Devine, "Dear Mr. Keppler," January 11, 1984.

28. Ibid.

29. Ibid.

30. City Council, "Resolution No. R/21-1984;" Ohio Consumers' Counsel, "Zimmer: The $3.5 Billion Question," Marian A. Spencer Papers, MSS 888, Box 27, Folder 23, William H. Zimmer Nuclear Power Station: Council Resolutions, 1983–84, CMC; Brewster Rhoads, "Statement by Brewster Rhoads, OPIC Director," January 6, 1984, Marian A. Spencer Papers, MSS 888, Box 27, Folder 30, William H. Zimmer Nuclear Power Station: Miscellaneous Reports, 1982–84, CMC.

31. Randy Ludlow, "For Dickhoner, Decision Means Worst Is Over," *Cincinnati Post*, January 23, 1984.

32. "Zimmer Plant Switches to Coal," *Cincinnati Enquirer*, January 24, 1984; CG&E, "Annual Report," 1984, DE; "Zimmer: A Troubled Past, an Uncharted Future," *Cincinnati Post*, January 23, 1984.

33. "Zimmer: A Troubled Past, an Uncharted Future," *Cincinnati Post*, January 23, 1984; Howard Wilkinson, "Zimmer Plant Switches to Coal," *Cincinnati Enquirer*, January 22, 1984.

34. Chew, "Power from the Valley," 142–150; Megan Chew, "The Ohio Valley's Nuclear Moment: Marble Hill and Madison, Indiana," *Ohio Valley History* 17, no. 1 (Spring, 2017), 29–51; Wellock, *Safe Enough*, 78; Casey Bukro, "Midwest a Graveyard for Nuclear Power Plants," *Chicago Tribune*, August 28, 1984; James Cook, "Nuclear Follies," *Forbes Magazine*, February 11, 1985.

35. See Bedford, *Seabrook Station*; and Aron, *Licensed to Kill*.

36. Author's interview with Phil Amadon, June 25, 2014.

Epilogue

1. CG&E, "Annual Report," 1984–90, DE; author's interview with Stan Hedeen, October 23, 2019.

2. Ohio Consumers' Counsel, "Summary of Utilities' Decision to Convert to Coal," January 23, 1984, Marian A. Spencer Papers, MSS 888, Box 27, Folder 31, William H. Zimmer Nuclear Power Station: Ohio Office of the Consumers' Counsel, 1984, CMC; "CG&E Auctions Zimmer Equipment," *Cincinnati Post*, August 19, 1985; David Ivanovich, "CG&E Settles Zimmer Lawsuit," *Cincinnati Post*, November 21, 1987; Bob Musselman, "Zimmer Designer, CG&E Settle Suit," *Cincinnati Post*, November 14, 1987; James F. McCarty and Jackie Jadrnak, "Customers Will Pay Half of Zimmer's Costs," *Cincinnati Enquirer*, November 27, 1985; James F. McCarty, "PUCO Listens to Pros, Cons of Zimmer Plant Settlement," *Cincinnati Enquirer*, November 7, 1985.

3. Jeff McKinney, "New Plan to Boost CG&E's Power," *Cincinnati Enquirer*, July 31, 1990.

4. Keith BieryGolick, "The Death of a Power Plant, Maybe a Village," *Cincinnati Enquirer*, July 10, 2022.

5. Richard Gibeau, "Neighbors Fight CG&E Ash Storage Site," *Cincinnati Post*, February 5, 1985; Richard Gibeau, "Groups Challenge Zimmer Conversion," *Cincinnati Post*, November 14, 1986; "Zimmer Safety Citations Set OSHA Record," *Cincinnati Post*, July 4, 1990; Len Penix, "Zimmer Plant Contractors Face Citations," *Cincinnati Post*, July 6, 1990.

6. Author's interview with David Fankhauser, August 5, 2017.

Index

Index